Saints and Sinners

The Latin Musical Dialogue in the Seventeenth Century

FRITS NOSKE

CLARENDON PRESS · OXFORD
1992

Oxford University Press, Walton Street, Oxford OX2 6DP
Oxford New York Toronto
Delhi Bombay Calcutta Madras Karachi
Petaling Jaya Singapore Hong Kong Tokyo
Nairobi Dar es Salaam Cape Town
Melbourne Auckland
and associated companies in
Berlin Ibadan

Oxford is a trade mark of Oxford University Press

Published in the United States
by Oxford University Press, New York

British Library Cataloguing in Publication Data
Data available
ISBN 0–19–816298–7

Library of Congress Cataloging in Publication Data
Data available
ISBN 0–19–816298–7

Set by Hope Services (Abingdon) Ltd
Printed in Great Britain by
Bookcraft (Bath) Ltd
Midsomer Norton, Avon

10000 73447

The music in Part Two is published with financial support
of The Netherlands Organization of Scientific Research.
Music origination by Per Hartmann.

For Michael Talbot
in friendship and gratitude

PREFACE

This book deals with a musical genre that has never been the object of comprehensive scholarly treatment. When, about eight years ago, I was doing research on Latin sacred music in the Dutch Republic, I came across a dozen pieces entitled *Dialogus* which aroused my interest. Consequently, I looked for literature on this type of composition. To my great astonishment, I found only a single, pioneering, article written by Howard Smither in 1967. This discussed the Latin dialogue in Italy during the first three decades of the seventeenth century. As for the dialogue repertory after 1630, this was completely *terra incognita*.

My project to write a study about this apparently neglected musical genre entailed extensive research, in particular in Italy, the country which produced about 90 per cent of the entire repertory. Since only relatively few dialogues exist in a modern edition, the greater part of the works discussed in this book had to be transcribed from the original sources, a tiresome but also rewarding business. A transcriber follows the compositional process closely and quite often discovers details which would remain concealed to anyone studying a score ready for use. The total number of dialogues that were available to me runs to more than 250—quite sufficient for a comprehensive description of the genre. While it is possible that important individual works have escaped my attention, it seems unlikely that these would affect the overall historical picture.

The division of the material offered some problems. The more I became familiar with the genre, the less the obvious chronological treatment, stressing the purely musical development at the expense of other characteristics, proved satisfactory. After considerable deliberation, I decided to divide the Italian repertory on the basis of subject-matter (biblical and non-biblical, with subdivisions). This created the need to write an introductory chapter about the genre's general aspects, musical as well as extra-musical. Because of the relatively small number of dialogues written outside Italy, these have been grouped according to their country of origin. Finally, a detailed analysis of six dialogues dealing with the subject of the Sacrifice of Abraham is given in Chapter 5. Taken together, these works reflect more or less the genre's development during the seventeenth century.

An interesting by-product of this study was the discovery of the remarkable compositional qualities of a few so-called minor masters such as Chiara Margarita Cozzolani, Alessandro Della Ciaia, and Giovanni Antonio Grossi. In addition, little-known aspects of the output of more famous figures are brought to light. For instance: while Giovanni Legrenzi's high reputation in the fields of opera and instrumental music is common knowledge, it now appears that as a young man he excelled in the writings of sacred works. These four composers, together with six others, are represented with transcribed dialogues in Part Two of this volume, an indispensable complement to the discussion of the repertory in Part One.

The following list of friends, colleagues, and other persons to whom I am indebted for information and help is of considerable length even though incomplete. Although I have limited myself to a rather dry enumeration, I recognize with gratitude that in many cases the simple mention of a name hides generous assistance in solving problems of various kinds.

As with my previous books on Sweelinck and the Dutch motet I owe a great debt to the dedicatee of this study, Michael Talbot (Liverpool). His correction of my English was as usual accompanied by valuable critical remarks. Graham Dixon (London), Jerome Roche (Durham), and Jonathan Wainwright (Oxford) most graciously gave me extensive bibliographical information. The same is true of Gunther Morche (Heidelberg) and Barbara Przybyszewska-Jarmińska (Warsaw), both of whom sent me in addition photocopies of their transcriptions of several works. Among other colleagues who furnished data and documents are Zygmunt Szweykowsky (Cracow), Iginio Ettore (Lecce), Andrew Jones (Cambridge), Robert Kendrick (New York), and Günther Massenkeil (Bonn). My sojourn in the Collegio Internazionale dei Sacerdoti del Sacro Cuore, both in Bologna and in Rome, was made possible by the late Monsignore Raffaele Forni (Lugano). Nor did I lack help in my local neighbourhood: Padre Ugo Orelli (Convento Capuccini, Faido) most kindly lent me for a long time a much-needed seventeenth-century *Concordantia bibliorum*, and I owe to Palma Forni (Villa Bedretto) precious information about locally venerated saints in Ticino and northern Lombardy. The number of librarians, archivists, and other scholars who gave me assistance far beyond what may reasonably be expected includes Giorgio Piombini (Civico Museo Bibliografico Musicale, Bologna), Lieuwe Tamminga (Archivio di San Petronio, Bologna), David Bryant (Fondazione Giorgio Cini, Venice), Alessandro Picchi (Archivio del Duomo, Como), Norbert Dubowy (Istituto Storico Germanico, Rome), Gertraut Haberkamp (Bayerische Staatsbibliothek, Munich), and the assistant archivist of the Fabbrica del Duomo in Milan, who wishes to remain anonymous. Finally I must mention with special gratitude the undeniable expert in the field of the sacred dialogue, Howard Smither (Chapel Hill, NC), who read part of the manuscript and offered critical observations.

This book is but a first step towards a broader knowledge about an important aspect of seventeenth-century sacred music. Much remains to be done, particularly in regard to the paraliturgical, religious, and social functions of the Latin dialogue. I cherish the hope that my effort will result in further studies answering questions that are still unresolved.

Frits R. Noske

Airolo (Ticino)
May 1991

CONTENTS

PART ONE

Historical and Analytical Observations

1

General Aspects

The Role Dialogue

The musical genre discussed in this book is that of the Latin role dialogue, a sung conversation on a sacred subject involving two or more characters, each of whom is represented by a single voice. Within this concept ensembles are employed—at least in principle—in a strictly realistic manner to convey the utterances of groups of characters. Role dialogues should be distinguished from other types, such as the merely verbal dialogue set to music in a manner which does not differentiate between the individual characters, or the instrumental dialogue, which, although presenting 'roles', lacks the characters. These genres are excluded from the present study, and the same is true of the 'solo' dialogue, a composition in which a single voice switches from one character to another.[1]

Thus, the role dialogue (henceforth simply called 'dialogue') is written, verbally as well as musically, in direct speech. The characters exchange statements or questions and answers, and eventually arrive at an understanding or at least a clarification of the situation. However, since the subject-matter is often based on a scriptural story or episode, the text may also contain portions written in indirect speech, sometimes indispensable for the comprehension of the sequence of events. These fragments are assigned either to a solo voice, which adopts a narrative role, detached from those of the characters involved in the scene, or to an ensemble, ranging in size from two to eight voices and representing an individual narrator. Both procedures can be applied within a single work.[2] A feature common to the greater part of this repertory is the moralizing statement or 'Alleluia' that forms the concluding section of the composition. Here the characters step out of their roles and address themselves directly to the listener. As with the set words of a group of persons and the ensembles representing the narrator, such a *conclusio* can provide a welcome contrast to the previous unrelieved alternation of solo passages, which is inherent in the genre. Moreover, it supplied a musical climax to the composition, the full complement of voices and instruments being involved.

---

[1] See, for instance, G. Carissimi, 'Lucifer caelestis olim'.

[2] The application of this procedure is found in several dialogues by L. Ratti, *Sacrae modulationes* (Venice, 1628), and P. Quagliati, *Motetti e dialoghi concertati, libro secondo* (Rome, 1627). Three of Ratti's dialogues are available in a modern edition.

A small number of dialogues do not conform to the stated criteria. They belong to two different types. First, there is the non-conversational dialogue, in which the characters abstain from reacting to each other's words; instead, they express alternately their views and feelings about the subject-matter, either individually or collectively. This kind of composition represents, so to speak, the musical counterpart of a *tableau vivant*, lacking any narrative or dramatic development.[3] The second type is that of the 'aborted' dialogue, a piece which opens with a conversational section but subsequently turns into a motet, the words being common to all the voices. It may be assumed that such anticlimactic texts were penned by authors who were prepared to sacrifice a balanced form for the sake of the content.[4]

Like opera, the dialogue owes its existence to the birth of monody in the last years of the sixteenth century. Although as a whole a dialogue cannot be a monodic composition, the alternation of solo voices presupposes the use of this new medium. Another precondition was the emergence of the *stile concertato*. Unlike the motet, which in principle could also be set in the *stile antico*, the dialogue was invariably written as a *concertato*, a term that implies some or all of the following characteristics: the interplay of voices with or without instruments; non-overlapping cadences; the polarity of the outer voices; the affective representation of words; a wide melodic range (especially in the vocal bass parts); free declamation alternating with dance-like rhythms; and the use of the thorough-bass as a *fondamento*. The application of these devices not only heightened the dramatic impact but also prevented the dialogue from adopting stereotyped schemes such as one encounters in opera, cantata, and various instrumental genres from the mid-century onwards.

The Religious Function

Why did composers write dialogues? There may be more than one answer to this question, but I believe that the main driving force was their endeavour to replace the past by the present. The textual subjects are either precisely fixed in time but very remote (stories and episodes from the Bible; hagiography) or timeless (abstract moralizing texts). In either case, a merely descriptive mode of presentation would fail to draw the listener into the scene; the temporal distance or the abstract content forms a psychological barrier. If, however, the scene is 'enacted' instead of related, remoteness and abstractness are transcended and

---

[3] An example of this type is Domenico Mazzocchi's *Dialogo del'Apocalisse*.

[4] An example of an aborted dialogue is found in the motet collection *Encomia sacra musice decantanda*, Op. 6, by the Dutch Carmelite friar Benedictus a Sancto Josepho (Utrecht, 1683). This is 'Audite virgines', which opens with an arresting exchange of speech between Jesus and two virgins, but then turns into a lyrical description by all three voices of the joyful unification with Christ, including an allusion to St Barbara. Unlike Mazzocchi (see n. 3), Benedictus did not call his composition a dialogue.

the desired effect is produced more easily. An analogous procedure is found in eighteenth-century epistolary fiction. In the novels of Richardson, Fanny Burney, and Laclos it is not only the addressee who receives the letter but also the reader; he or she gains the impression that the events are taking place here and now. In more or less the same way, the person listening to a dialogue feels that somehow the 'actors' are addressing him or her as well as each other.

Throughout the ages dialogued texts have served instructional and educational purposes. Philosophers from Plato onwards presented their ideas in the form of arguments and counter-arguments assigned to different persons. The same is true of instruction on a lower level. As late as the seventeenth century, schoolbooks were often styled as extended dialogues between master and pupil. Since the Latin musical dialogue deals with sacred subjects, the Church and various religious orders accepted the cultivation of the genre, not least because of its educational and edifying character. This raises another question: what, exactly, was the function of the dialogue within the ambit of the Church?

Textbooks dealing with the history of music rarely mention the seventeenth-century dialogue. This may be the reason why an unhappily phrased statement by Manfred Bukofzer has carried more weight than it deserves. His assertion that 'Latin dialogues . . . are strictly liturgical' is misleading indeed, especially because of the inclusion of the word 'strictly'.[5] The (perhaps unintended) implication is that dialogue texts belong to the official and universal Roman liturgy. In this sense Bukofzer's statement is decidedly wrong. Practically no dialogue was set to a purely liturgical text, and even scriptural fragments involving verbal interchange were only rarely presented without variation, abbreviation, or amplification. If, however, we read the term 'liturgical' as 'used within the liturgy' or, by extension, 'during the service', then Bukofzer is right. Although Latin dialogues may have served private or public devotions, they were written primarily for performance in church. Many of them are explicitly assigned to specific feast-days; this is particularly true of compositions celebrating saints.

So far, it has been possible to determine a position within the church service for only a few dialogues. This is the case with four pieces by Lorenzo Ratti whose titles mention their use as substitutes for the Offertory in the Mass on specified days.[6] To these I can add two more examples: Carlo Milanuzzi wrote dialogues for the feasts of St Charles Borromeo (4 November) and St Stephen (26 December). Both these works ('Deus qui vides' and 'Exaudi, Domine') are entitled 'Introductio ad Vesperas'.

Unfortunately, this is where our knowledge ends. Therefore most of the

---

[5] M. Bukofzer, *Music in the Baroque Era* (New York, 1947), 124.

[6] See Ratti, *Sacrae modulationes*. The four dialogues are 'Cum descendisset Jesus', 'Erat quidam regulus', 'Homo quidam fecit caenam magnam', and 'Simile est regnum caelorum'. Regina Chauvin's assertion that these pieces 'are the only 17th-century dialogues to have served a liturgical function' is, of course, untenable (*New Grove* (London, 1980), xv. 600).

following observations are necessarily undocumented. Yet they are supported by significant circumstantial evidence.

In the seventeenth century it was customary to use motets or instrumental pieces as substitutes for items from the Proper of the Mass and Vespers. In addition, motets or *sonate da chiesa* could be employed as introductory music. The functional indications heading the above-mentioned compositions by Ratti and Milanuzzi show that dialogues were included in this practice. Apart from the Offertory, the Gradual could be replaced by a motet or a dialogue, and the same is true of the repeats of antiphons after the psalms and the Magnificat at Vespers. This freedom was frowned upon by the ecclesiastical authorities in Rome and even formally condemned during the pontificate of Alexander VII. Two papal decrees, issued in 1657 and 1665, stated that only texts contained in the Missal and Breviary should be sung in the Mass and Divine Office. Moreover, it was strictly forbidden to set texts from liturgical books, the Scriptures, and the church fathers in any way other than their original form.[7] However, despite the fact that the *maestri di cappella* who overstepped these rules were threatened with ignominious dismissal, the decrees were virtually ignored all over Italy. This was even true of a composer who lived and worked in close proximity to the Vatican: Giacomo Carissimi.

Yet in other respects the Church adopted a liberal attitude. It allowed the celebration of local saints, some of whom were hardly known outside their respective cities. This practice, sanctioned by the Council of Trent, weakened to a certain extent the dogma of universality that was strongly favoured during the Age of Absolutism, the more so as these local feasts often assumed greater importance than the feasts celebrated universally in Roman Catholic christendom. Gasparo Casati, a composer working in the north Italian city of Novara, clearly exemplifies this parochial attitude. His dialogue 'Quam laetam hodie videbo hanc civitatem' (1640) was written in honour of the local patron, San Gaudenzio. In this piece a stranger marvels at the festive atmosphere, the joy of the people, and the crowd hurrying to church. Another character, obviously a citizen, explains the reason to him and, at the stranger's request, enumerates all the benefits the saint has conferred on the city. The text includes words like 'Novarienses cives' and 'urbs nostra', stressing the local ambience of the scene.

In addition to works written for services in cathedrals, collegiate churches, and small parish churches, some dialogues were composed for institutions on the periphery of the church, such as seminaries, convents, secondary schools run by religious orders, academies specializing in sacred music, and conservatories or *ospedali*. Very little is known about these practices, but we may safely assume that the extant dialogues by Lorenzo Ratti and Giacomo Carissimi, successive *maestri di cappella* at the famous Roman seminary, the Collegium

---

[7] See A. V. Jones, *The Motets of Carissimi* (2 vols.; Ann Arbor, Mich., 1982), i. 143 ff.

Germanicum, were intended for the associated and adjacent San Apollinare. The predecessors of these two composers—Agostino Agazzari, Antonio Cifra, Ottavio Catalani, and Annibale Orgas—may also have contributed dialogued music for the rites in this church.

Since a number of composers were monks or nuns, their dialogues may well have served monastic practice. In the case of Chiara Margarita Cozzolani, a Benedictine nun from the Milan convent of Santa Radegonda, this is more than likely. One of her printed collections of sacred music includes a Vesper psalm 'Beatus vir' transformed into a dialogue (1650). This work would have been quite unacceptable for the celebration of Divine Office in church.

In view of the genre's educational character, it is hardly mere speculation to imagine that the dialogue was used for religious instruction in schools within Italy and elsewhere. Details are as yet unknown, but the fact that Carlo Donato Cossoni left his unpublished manuscripts, including two dialogues, to the Collegio in the Swiss town of Bellinzona, may be significant. This school was run by the Benedictines; today the manuscripts are in the library of the Einsiedeln monastery.

Among the numerous academies in Italy a few specialized in sacred music: for instance, the Accademia della Morte and the Accademia dello Santa Spirito in Ferrara. These institutions possessed small musical establishments of their own, directed by renowned musicians such as Alessandro Grandi, Maurizio Cazzati, and Giovanni Legrenzi. Grandi and Legrenzi especially wrote a number of motets and dialogues for these charitable confraternities. However, none of this removes the possibility that these pieces were sung during church services.

Unlike schools managed by members of religious orders, Italian conservatories and *ospedali* were mostly administrated and controlled by laymen, that is, nobles in Venice and confraternities in Naples. Yet in both cities the participation of the students in church ceremonies was virtually taken for granted. As with monasteries and schools, specific knowledge about the function of our genre in these institutions is lacking. The Venetian composer Natale Monferrato, who was not only *maestro di cappella* at St Mark's but also musical director of the Ospedale dei Mendicanti, may have written his motets and dialogues for this institution rather than the ducal church, for whose rites he favoured music in the *stile antico*.[8]

It seems that in Italy music explicitly intended for private devotions was practically confined to collections of solo motets. Examples can be found in Orazio Tarditi's *Celesti fiori musicali*, op. 8, and similar books by him (opp. 23 and 26), all of which mention, alongside other instruments, the harpsichord as the bearer of the continuo part. However, it is quite possible that individual small-scale motets and dialogues taken from the numerous collections printed during the

[8] See D. Arnold, 'Monferrato, N.', *New Grove* (London, 1980), xii. 481–2.

seventeenth century also served this purpose. Although written primarily for the church, they may have been performed domestically as well.

The term 'domestic' may be expanded to include the private musical establishments of princes and cardinals. One of the largest bodies of this kind was that of Marie de Lorraine, the duchess of Guise, in Paris. Her *maître de chapelle*, Marc-Antoine Charpentier, wrote several dialogues during the years of his employment by this distinguished noblewoman. Leaving aside the chapels of reigning princes, similar establishments were rare or non-existent in Protestant regions of Europe. On the other hand, the practice of observing private devotions in the houses of individual citizens seems to have been more widespread than in Catholic countries. In the previous century both Luther and Calvin had strongly recommended this kind of religious activity, terming it 'häusliche Andacht' and 'exercice spirituelle' respectively. While it is true that the dialogues and motets by the Dutch composers Jan-Baptist Verrijt and Benedictus a Sancto Josepho were intended for the service in Catholic churches, these same pieces were actually performed in Protestant domestic circles, as well as during the sessions of the local *collegia musica*.[9]

Since the Lutheran Church lacked a central authority, its mode of worship was pluriform and flexible; during the sixteenth and seventeenth centuries it changed constantly. This permitted a wide musical scope. Although dialogued texts in the vernacular became ever more predominant (as, for example, in the scriptural *historia*), the use of Latin was never abolished. Latin motets and dialogues may have replaced items from the Proper of the Mass (removed by Luther on dogmatic grounds) or served as illustrations to sermons. Kaspar Förster, the most prolific German composer of Latin dialogues, was a Catholic. Having studied with Marco Scacchi in Warsaw and Carissimi in Rome, he nevertheless became employed as royal *Kapellmeister* at the court of the Danish king Frederick III, serving in this capacity from 1652 to 1655 and 1661 to 1667. Although none of his dialogues is dated, there are indications that they were written for Lutheran services during the composer's second stay in Copenhagen. In any case the copious use of instruments points to Denmark rather than to Italy, where, at least in the mid-century, the inclusion of instrumental parts in the dialogue was exceptional.

From the above observations it appears that the religious functions of the dialogue paralleled those of the motet. For the rest, many questions have been raised and only a few answered. Much research has still to be done to clarify and complete the picture.[10]

---

[9] S. Spellers, 'Collegium Musicum te Groningen', in *Bouwsteenen: Derde jaarboek der Vereeniging voor Noord-Nederlands Muziekgeschiedenis* (n.p., 1881), 22–9; F. R. Noske, *Music Bridging Divided Religions: The Motet in the Seventeenth-Century Dutch Republic* (2 vols.; Wilhelmshaven, 1989), i. 22–5.

[10] The function of the dialogue in the oratory will be discussed in the next section of this chapter.

Delimitation

If we take the dialogue as a separate kind of composition, it should be delimited from neighbouring genres, synchronically as well as diachronically. To draw a line between polyphonic dialogue motets from the late Renaissance and role dialogues from the early Baroque is rather easy. In the first genre the dialogue is almost exclusively verbal, whereas in the second both text and music convey the exchange of speech between individual characters or groups. This can be demonstrated clearly by comparing a number of Renaissance motets with a Baroque dialogue written on the same words. The paraphrased Annunciation text 'Missus est Gabriel Angelus' was set twice by Josquin des Prez (a 4 and a 5), later by Orlandus Lassus (a 6), and after the turn of the century by the Florentine Severo Bonini (for two solo voices, a five-part 'choro', and continuo). Josquin does not make any attempt to differentiate between the characters involved, that is, the archangel, the Virgin, and the narrator. The same is true of Lassus, although in his piece the narrator is slightly profiled by the use of non-overlapping cadences. In each of the three sixteenth-century motets the characters speak by means of all available voices, singing in a polyphonic style. Bonini's scoring, on the other hand, clearly distinguishes between the different roles: a countertenor part is assigned to the archangel, a soprano part to the Virgin, and a chorus (probably meant to be an ensemble of solo voices) to the narrator. The fact that the voices of Gabriel and Mary are both absorbed by the chorus does not disturb the listener, since these characters lose their identity in the narrative sections.[11]

Thus the compositional technique of the Renaissance appears to be quite incompatible with the concept of the role dialogue. This is even true of a piece that is termed 'concerto'. The motet 'Adam, ubi es?' by the Lucca organist Jacopo Corfini, included in his *Concerti continenti musica da chiesa* (Venice, 1591), deals with the expulsion from Paradise. God and Adam each speak by means of a separate four-part ensemble. However, in the course of the composition the voices previously representing Adam shift to Eve, and finally all eight parts sing God's verdict addressed to the serpent.

The first steps towards the assignment of a role to a single voice were taken by Gabriele Fattorini in a dialogue about the Resurrection involving Mary Magdalene and Peter. This work dates from 1600. Two years later Lodovico Grossi da Viadana made a similar effort in a piece dealing with the Finding of Jesus in the Temple. Although neither Fattorini nor Viadana was yet wholly consistent in coupling special voices to individual speakers, we may consider these two works as the beginning of the era of the role dialogue.

---

[11] This comparative description is borrowed from John Whenham's article 'Dialogue, Sacred', *New Grove*, v. 419–20.

Italian and Latin dialogues, based on sacred subjects, have much in common. Yet they differ on two major points. While the Latin texts are almost exclusively written in prose, the Italian words are always fashioned into verses, as a result of which they show less dependence on the scriptural readings. Even more important is the functional distinction. Unlike Latin dialogues, the Italian ones were used exclusively for private or public devotions outside the church. The functional differences between Latin dialogues and those written in the vernacular did not exist in Protestant regions of Europe such as northern Germany. But there the latter kind far outnumbered the former. In addition to both Catholic and Protestant Germany, England produced a number of sacred dialogues in the vernacular. Like the German pieces, these have been the object of much research. On the other hand, the Italian dialogue, cultivated by Antonio Cifra, Giovanni Francesco Anerio, Paolo Quagliati, Pietro Paolo Sabatini, and many other composers, still awaits a comprehensive scholarly investigation.

Although throughout the century Latin dialogues were considered in a loose sense motets, the differences between the two genres are evident. The dialogue offers more possibilities of dramatic poignancy. The motet, on the other hand, being less dependent on the text, has more musical resources at its disposal. It was probably for this reason that several composers wrote semi-dialogues, sometimes called 'motetto dialogato' or 'motetto in modo di dialogo'. The motet 'Morior misera' (1665) by Carlo Donato Cossoni lacks such a title; nevertheless this piece, dealing with a soul in distress, offers a beautiful example of this type of composition. The first part of the text is set homophonically for three voices over a chromatically descending bass emblematic of the traditional *lamento*. In the course of the piece this section is repeated three times and thus serves as a vocal ritornello. Between the repeats, various solo voices address the 'anima dolente', pitying her but also showing her the path to salvation. This motet cannot be described as a true role dialogue; yet the work possesses various traits of the genre.

Despite the thorough research of Howard Smither and Graham Dixon, the musical distinction between dialogue and early oratorio poses a problem that seems unsolvable.[12] Seventeenth-century terminology is anything but helpful. Small pieces were sometimes called 'oratorio', while extended compositions, involving a choir alongside solo voices, were designated 'dialogo'.[13] The matter

---

[12] H. E. Smither, 'The Latin Dramatic Dialogue and the Nascent Oratorio', *Journal of the American Musicological Society*, 20 (1967), 403–33; id., 'What is an Oratorio in Mid-Seventeenth-Century Italy?', *International Musicological Society Congress Report*, xi (Copenhagen, 1972), 657–63; id., 'Carissimi's Latin Oratorios: Their Terminology, Functions, and Position in Oratorio History', *Analecta musicologica*, 17 (1976), 54–78; id., *A History of the Oratorio*, i. *The Oratorio in the Baroque Era: Italy, Vienna, Paris* (Chapel Hill, NC, 1977); G. Dixon, 'Oratorio o motetto? Alcuni riflessioni sulla classificazione della musica sacra del seicento', *Nuova rivista musicale italiana*, 17 (1983), 203–22.

[13] An example of a particularly short composition is *Tobiae Oratorium* by Francesco Foggia (duration *c*.6 minutes). On the other hand, Carissimi's *Jephte* (duration *c*.25 minutes) is called 'dialogus' by Athanasius Kircher in his *Musurgia universalis* (Rome, 1650), i. 603.

is complicated further by the fact that the word *oratorio* was used not only for the building or the music, but also occasionally for the devout gathering. As far as we know, Latin oratorios of the seventeenth century were performed only at the S. Marcello oratory of the 'Arciconfraternità del Santissimo Crocifisso' in Rome, and even this practice was restricted to the Lenten period. This would mean that the compositions by Carissimi and Mazzocchi on the subject of the Resurrection cannot be called oratorios, since they were written for the service on Easter Sunday. The assumption that the building of the SS. Crocifisso confraternity was the only place where Latin oratorios were performed seems to be contradicted by the titles of Mazzocchi's *Sacrae concertationes* and Graziani's 'quattro mute di dialoghi', which, although announced in the latter's double-choir psalms (op. 17) and Masses (op. 18), were almost certainly never published. It is puzzling indeed that both composers wrote these works 'pro oratoriis' and 'per gli oratorij' respectively, using the plural. However, it cannot be excluded that in these cases the meaning of the word *oratorio* is that of 'devout gathering'. Although we know that there was a clear functional difference between dialogue and early oratorio, this difference cannot be detected either from the music or from the text. (Only in the case of a text unsuited to the Lenten period can we be sure that it was written for a non-oratorical dialogue.) Hence I have decided arbitrarily to draw a line at the length of five hundred bars, taking the semibreve as a bar-unit in common time. Compositions exceeding this number are not discussed in the present study. On the other hand, the relatively short works intended to be sung in the SS. Crocifisso oratory will be termed 'oratorio dialogue'.

Sources and Composers

Printed dialogues were practically always included in motet collections. Books containing dialogues alone are very rare. Giovanni Francesco Capello published such a volume in 1613. The four collections of oratorio dialogues by Bonifazio Graziani, already mentioned, have not survived. It was only around 1700, when the genre was almost extinct, that two more books of dialogues appeared in print: the *Sacri dialoghi*, op. 1, and the *Dialoghi sacri*, without opus number, by Ippolito Ghezzi. Some motet collections mention the inclusion of dialogues on their title-pages; this is the case with certain works by Rognoni, Fergusio, Agazzari, Capello, Quagliati, Banchieri, Bazzino, Sister Cozzolani, and Pfleger.[14]

---

[14] F. Rognoni Taeggio, *Missa, salmi intieri e spezziati, Magnificat, falsibordoni, motetti con 2 dialoghi* (Milan, 1610); G. B. Fergusio, *Motetti e dialoghi per concertar* (Venice, 1612); A. Agazzari, *Dialogici concentus*, Op. 16 (Venice, 1613); G. F. Capello, *Motetti e dialoghi*, Op. 7 (Venice, 1615; P. Quagliati, *Motetti e dialoghi* (Rome, 1620); id., *Motetti e dialoghi concertati*, libro secondo (Rome, 1627); A. Banchieri, *Dialoghi, concerti, sinfonie e canzoni*, Op. 48 (Venice, 1625; 2nd enlarged edn., Venice, 1629); N. Bazzino, *Motetti et dialogi* (Venice, 1628); C. M. Cozzolani, *Salmi . . . motetti e dialoghi*, Op. 3 [= Op. 4] (Venice, 1650); A. Pfleger, *Psalmi, dialogi et motetti*, Op. 1 (Hamburg, 1661).

In a considerable number of books the index distinguishes between the motets proper and the dialogues. However, many publications do not provide even this information: the word 'dialogue' appears neither on the title-page nor in the index—nor even at the head of the individual pieces. This may be the main reason why the genre was for a long time not recognized as such. To track down dialogues is indeed difficult, the more so as the identity of the characters is not always clearly established. In quite a few cases the names of the 'actors' have to be derived indirectly from the context, and even then it often remains unclear whether God, Christ, or an angel is speaking.

Alongside the printed works, many dialogues are preserved in manuscripts. These are found mostly in Italian church archives. The largest collection of this kind is that of Giovanni Antonio Grossi, who worked at the cathedrals of Crema, Piacenza, and Novara, and subsequently at the church of S. Antonio in Milan. In 1669 he won the competition for the office of *maestro di cappella* at Milan Cathedral, a post which he held until his death in 1684. Since Grossi, unlike many other composers, used to keep the manuscripts of his own works in his possession, his enormous output of some five hundred sacred compositions, including about sixty dialogues, is now in the Cathedral Archives of Milan.

According to a rough estimate, about 90 per cent of the dialogue repertory was composed by Italians. Although the most famous Italian composer, Claudio Monteverdi, did not leave any true Latin dialogue, many masters of high repute contributed to the genre: Agostino Agazzari, Giovanni Francesco Anerio, Adriano Banchieri, Stefano Bernardi, Giacomo Carissimi, Gasparo Casati, Maurizio Cazzati, Giovanni Paolo Colonna, Ignazio Donati, Francesco Foggia, Alessandro Grandi, Bonifazio Grazziani, Sigismondo d'India, Giovanni Legrenzi, Marco Marazzoli, Domenico Massenzio, Domenico and Virgilio Mazzocchi, Tarquinio Merula, Natale Monferrato, Paolo Quagliati, Lorenzo Ratti, and Orazio Tarditi—not to speak of numerous 'minor' musicians. A great number of these were priests; their clerical status probably facilitated their careers. As for the authors of dialogue texts, they are virtually unknown. It may be assumed that composers often wrote their own texts. Yet one comes across a few works whose words do not lend themselves easily to the composition of an effective dialogue. Perhaps in these cases the authors were other members of the clergy who were rather ignorant in matters of music.

A great portion of the repertory written by Italian composers is found in the Civico Museo Bibliografico Musicale in Bologna. Other dialogues are preserved in libraries and archives in Venice, Milan, Ferrara, Lucca, Rome, Naples, and elsewhere in the Italian peninsula. As for sources outside Italy, libraries in Munich, Regensburg, Frankfurt, Münster, Wrocław (Breslau), Paris, London, and Oxford (Christ Church College) should be mentioned in particular. Modern editions are as yet rare. Apart from a few pieces published independently, there are only two collections. One, edited by Howard Smither, contains twenty-seven dialogues by various composers from the period 1600–30.[15] The other, an edition by Wolfgang Witzenmann of Mazzocchi's *Sacrae concertationes* (1664), includes eight dialoghi; all but two of these are oratorio dialogues.[16]

The Texts

The authors of dialogued texts based on the Scriptures were faced with two problems. The first concerns the verbal configuration. The biblical words, abstracted from a continuous narrative discourse, almost never present a rounded form suited to a musical setting. This difficulty is particularly evident in the final part of a scriptural dialogue lacking a narrative role. Whereas in the Bible only one person speaks at a time, the musical composition must be concluded with an ensemble involving all the voices. It was left to the composers to deal with this problem; however, at first their solutions turned out far from satisfactory. Viadana's 'Fili, quid fecisti', mentioned earlier in this chapter, opens with a complete dialogue between Mary, Joseph, and the twelve-year-old Jesus.[17] Subsequently the entire text is recapitulated, this time set homophonically for all three voices. The result is rather uncomfortable: a dialogue and a motet using the same verbal material are juxtaposed. Luigi Aloysius Balbi also treated the episode of the Finding of Jesus in the Temple (1606); unlike Viadana, he adhered strictly to the biblical text (Luke 2: 48–9), that is, with spoken roles for the Virgin and Jesus only. In this piece the last three words of Jesus ('quia in his quae Patris mei sunt *oportet me esse*') are sung by both voices. If the final phrase is allowed to assume the character of a doctrinal statement, its setting for more than one voice will prove less disturbing. This happens, for example, in Banchieri's dialogue (1625) dealing with Jesus and Peter: 'Tu es Petrus et super hanc petram *aedificabo Ecclesiam meam*' (Matt. 16: 18).[18] Yet in all these cases at least one character exchanges his identity for that of another. It is, therefore, hardly astonishing that in the course of the second decade of the century this procedure was gradually abandoned. Instead, a more satisfactory solution was adopted: the addition of a final section to the dialogue proper; this was set either to narrative scriptural words or to a freely invented text. Several examples of this practice are contained in Smither's anthology (Grandi, Massenzio, Vincenzo Pace, Donati, Anerio, Banchieri, Bazzino, and Ratti).

The second problem originated from the sober, almost laconic, wording of the scriptural episodes, offering insufficient verbal material for the presentation of a true 'scene'. This induced the authors to elaborate the textual details. 'Heu!

---

[15] *Antecedents of the Oratorio: Sacred Dramatic Dialogues, 1600–1630*, ed. H. E. Smither (Concertus Musicus, 7; Laaber, 1985), henceforth abbreviated as Smither *SDD*.

[16] C. Mazzocchi, *Sacrae concertationes*, ed. W. Witzenmann (Concertus Musicus, 3; Cologne, 1975).

[17] The text is borrowed from the Magnificat antiphon of the Second Vespers on the day of the Holy Family.

[18] A. Banchieri, 'Petre, amas me?'

Domine, respice et vide' by Giovanni Antonio Grossi is a dialogue on the subject
of the announcement of the forthcoming birth of Isaac, which, overheard by
Sarah, causes her to laugh.[19] When God rebukes her for this, Abraham's wife,
being afraid, denies the charge. But God insists: 'Nay, but thou didst laugh.' Such
is the biblical text (Gen. 18: 1–15). In the dialogue it becomes three angels, representing God, who reprove Sarah. She reinforces her original denial ('minime'),
whereupon the angels shout in chorus 'Non! maxime!' The addition of this
humorous detail enlivens the scene greatly.

The cultivation of individual details sometimes led to the writing of a text
which overlaid the original scriptural words. 'Fili, ego Salomon', set by Pietro
Lappi (1614) as well as the Dutchman Jan-Baptist Verrijt (1649), is based on a
single biblical verse: 'My son, if sinners entice thee, consent thou not' (Prov. 1:
10). Starting from these few words, the author constructs a scene involving King
Solomon and his son. The latter admits and deeply deplores his choice of bad
company ('Heu! Pater peccavi'). The king tells him not to sin anymore but to
pray so that he will be forgiven. The dialogue proper is interrupted three times
by a short tutti section explicitly addressing the listener: 'Hear me and take my
words to heart;' the same ensemble concludes the piece with the words: 'Let us
pray, O ye sons, to Christ, King and Saviour, and, with a contrite heart, let us ask
for forgiveness.' Thus a single verse is transformed into a true scene, the didactic
character of which is evident. Solomon's original warning to an individual person (his son) has become an admonishment to all young men.

This dialogue belongs to the no man's land between scriptural and non-scrip-
tural texts; although derived from a biblical subject, the words express freely
invented thoughts and actions. Other dialogues, dealing with subjects that lie
outside the Scriptures, may also be assigned to this intermediate category, albeit
for opposite reasons. Some free texts are less free than they seem, since they
swarm with quotations from the Book of Psalms and the Canticles. An example is
'Vias tuas, Domine', set by an anonymous composer. The text of this dialogue
combines verses taken from Psalms 24 and 138 with various fragments from the
Song of Songs.

Completely free texts were used increasingly after 1630. These mostly treat
subjects such as that of a sinner to whom Christ or an angel shows the way
to salvation, a discussion between allegorical characters (for example,
personifications of the World, Heaven, and Hell), or the celebration of a saint.
Pieces belonging to the third category often features artificially dialogued texts:
questions are interpolated within a basically reflective or narrative discourse.
Here follows an example in which this procedure is taken to extremes. This is a
dialogue in honour of St Catherine of Siena, composed by Sister Chiara
Margarita Cozzolani (1650); the characters are three Homines and two Angeli.

---

[19] Transcription in Part Two.

HOMINES. O Caeli cives, o Angeli pacis, audite, volate, venite, narrate: ubi pascat, ubi cubet Christi Sponsa Catharina.

ANGELI. In caelo quiescit et inter Sanctos pax illius est.

HOMINES. O felix requies, beata sors! Dicite nobis: ubi regnat exaltata coronata Christi Sponsa Catharina?

ANGELI. In caelo nunc regnat et inter Sanctos regnum eius est.

HOMINES. O felix regnum, aeternum regnum, beata sors! Dicite nobis: ubi Regina gloriosa triumphat?

ANGELI. In caelo triumphat et inter Sanctos palma illius est.

HOMINES. O felix triumphus, o palma beata, beata sors! Dicite nobis: ubi iubilans gaudet, exultat, laetatur iocunda Catharina?

ANGELI. In caelo congaudet, exultat, laetatur, et gaudium eius plenum est.

HOMINES. O dulcis risus, o felix gaudium, beata sors! Ergo casta Christi Sponsa Catharina in caelo quiescit?

ANGELI. In aeternum.

HOMINES. In caelo nunc regnat?

ANGELI. In aeternum.

HOMINES. In caelo triumphat?

ANGELI. In aeternum.

HOMINES. In caelo laetatur?

ANGELI. In aeternum.

TUTTI. In aeternum, in caelo nunc regnat, quiescit, triumphat, exultat, laetatur, in aeternum cantabit Alleluia!

This text is, of course, totally undramatic. When read, the words almost give the impression of childishness, but the fact that they are sung makes all the difference. The Nativity antiphon 'Quem vidistis pastores' may have served as a model for this kind of artificial dialogue.

The numerous scenes involving a soul in distress offer little variety. One of the means to enliven and dramatize these texts was to assign a role to the devil. However, in practice this could lead to embarrassing situations. In a dialogue by Giovanni Antonio Grossi in which an angel and the devil fight for the soul of Job, the latter finally opts for the side of heaven.[20] Both he and the angel conclude the piece by singing 'Sit nomen Domini benedictum', words that are logically contradicted by the devil, who sings simultaneously 'Sit nomen Domini maledictum'. This is the reading of the separate parts. In the score, however, the devil shares the text of the angel and Job, an ostensible anomaly justified by the addition of the word 'nolens' (against my will). It is quite possible that the words were changed because the bass singer flatly refused to blaspheme God in church.

---

[20] G. A. Grossi, *Historia di Giobbe*. Despite a few quotations from the Book of Job, the text of this dialogue is freely invented. In the Scriptures there is no verbal intercourse between Job and an angel. Nor does either of them speak with Satan.

A number of 'free' dialogues contain refrain-like sections serving as a vocal ritornello. In most cases these comprise nothing but a simple Alleluia, modelled on that of the plainsong antiphon 'Regina caeli' and probably added by the composer, who not only needed a change from the continuous alternation of solo voices but also wished to cast his piece in a rondo-like form.

Although as a rule Latin dialogues were written in prose, after 1650 we encounter a few sections of text in rhymed verses. The following example is taken from the opening section of Giuseppe Allevi's 'Infelix anima' (1668), sung by the Angelo Custode:

Infelix anima, quae inimica caelo, amica vitiis, abundans scelere, onusta maculis, in sorde manes, cur hodie non properas, velox non advolas ad fontem gratiae, Reginam superum?

> Est Maria fons perennis
> gratiarum ac salutis.
> Accedentes replet bonis
> et exornat vitae donis.
> Advocatam semper habet
> qui cum spe Mariam adit,
> nec quis unquam frustra accedit,
> sed quod petit certo tenet.

The prose passage and the octosyllabic lines suggest a setting in semi-recitative and aria style respectively. This is exactly what Allevi does in his composition. Other dialogues including verse sections by Francesco Bagati[21] and Augustin Pfleger[22] apply the same formal procedure. Towards the end of the seventeenth century this alternation of prose and verse led to the fixed combination of true recitatives and arias, transforming the dialogue into a dramatic 'cantata' and marking the end of the genre's historical development.

Scoring

The number of vocal parts in dialogues varies from two to nine; the maximum number of solo voices representing individual characters is four. Ensembles employed for groups (shepherds, angels, demons) or the transmission of narrative text sections are sometimes referred to as 'choro'; this term denoted the combination of three or more single voices rather than a choir in the modern sense of the word. Occasional use of vocal ripieni cannot be entirely excluded but, in general, the small number of singers in Italian churches as well as those in other countries was quite insufficient for this practice. Only Marc-Antoine

---

[21] F. Bagati, 'O animae felices' (1658).   [22] A. Pfleger, 'Jesu, amor dulcissime' (1661).

Charpentier prescribed the participation of a true choir; however, this was mostly done in large-scale works that exceeded the scope of the dialogue as defined at the beginning of this chapter.

In Italy the soprano part was sung by either a boy or a castrato, and the alto by a falsettist; in northern and western Europe the falsettist also sang the highest part. Female voices, prohibited in church, could be employed only in convents and *ospedali* or conservatories. It is unlikely that women participated in works performed in the oratory of the SS Crocifisso (though Mazzocchi's dialogue *Lamento di David* includes a *chorus mulierum* for four sopranos).

The roles of females were generally taken by high voices and those of men by low voices. An angel, being sexless, could be represented by any voice; representation by a bass was exceptional, however. The few deviations from these principles include the part of Sarah in Grossi's 'Heu! Domine, respice et vide', which is assigned to a tenor; but this might be explained by the fact that she is ninety years old. Another example of unusual scoring occurs in the dialogue 'Cede Constantia' by the Como composer Francesco Rusca. Its 'cast' consists of Santa Costanza, a Tiranno, and three Assessori. As the martyr sings soprano and the tyrant bass, one would expect middle voices to be given to the assessors. However, this is not the case. All three of them are sopranos, weakening in this way the prominent part of the saint. Obviously tenors and countertenors were not at hand—for what reason one can only guess—and so the composer had to content himself with an unsatisfactory solution.

The Renaissance technique of representing single characters by groups of voices was not entirely abandoned during the early years of the seventeenth century. In his *dialogo per due chori*, 'Petre, amas me?' Leone Leoni presents the scene of Jesus and Peter (John 21: 15–17), each of whom is rendered by a four-part ensemble. Another double-choir work by this composer, 'Adjuro vos, filiae Jerusalem', adopts a slightly different form of 'casting', opposing a single female to a group of persons. Lodovico Bellanda applied this procedure on a smaller scale. In his four-part Canticle dialogue 'Surge, propera, amica mea' he represents a conversation between the Sponsus and the Sponsa by continuous alternation of two pairs of voices. The fundamental difference between the scoring of these pieces and that of the dialogued motet of the late Renaissance is that the ensembles keep strictly to their allotted roles. Unlike the vocal groups in Corfini's *concerto* (see earlier in this chapter), those of Leoni and Bellanda do not migrate from one character to another.

Leone Leoni and Paolo Quagliati, working in Vicenza and Rome respectively, published their double-choir dialogues with two continuo parts, each of which serves as *fondamento* of a four-voiced ensemble. This scoring points to performance in churches with more than one organ. Yet a continuo part could also be played on a plucked instrument. In the dialogue 'Intravit Jesus in quoddam castellum' by the German composer Daniel Bollius the voice of Maria Magdalena is supported by a figured bass written for the lute or theorbo; the part of the Historicus, on the other hand, has a continuo of its own, to be played on the organ.[23]

In Italy instrumental parts were rarely added to church dialogues. Among the few pieces of this kind are Capello's 'Abraham, Abraham!' (1615, with four unspecified instruments), Grancino's 'Quid est quod dilectus meus' (1631, with one violin), and Carissimi's 'Doleo et pœnitet me' (undated, with three viols).[24] Roman oratorio dialogues make more use of instruments. Although omitted from the publication, Domenico Mazzocchi prescribed in his *Sacrae concertationes* instrumental interludes and doubling of vocal ensembles. Instruments are also added to all but one of the five oratorio dialogues by Marco Marazzoli. As regards northern Italy, later in the century we find strings included in the scoring of dialogues by Francesco Petrobelli and Francesco Rusca. Yet most of the other composers (Monferrato, Urio, Sanromano, Bassani, Cossoni) still continued to score their works for voices and continuo alone.

Outside Italy the practice was entirely different, the use of concertizing instruments being the rule rather than the exception. Early examples are found in the dramatic works of Daniel Bollius (cornett, bassoon, lute). The three most important composers of dialogues among the *oltrimontani*, the Frenchman Marc-Antoine Charpentier, the cosmopolitan Kaspar Förster, and the Dutch friar Benedictus a Sancto Josepho, employed instruments in various ways: as concertato partners, as a means to achieve formal balance (*sinfonie*, ritornellos), and as added colour, more or less 'realizing' the continuo. The same is true of other composers, notably Anton Vermeeren (Antwerp), Henry Du Mont (Paris), and Bartłomiej Pękiel (Warsaw, Cracow). In addition to strings, wind instruments were used by Charpentier (recorders) and Benedictus (trombones and bassoon). While in general instruments had a purely musical function in the dialogue, they could also contribute to the scene's dramatic impact. This will be shown in Chapter 5.

Form and Structure

The formal layout of a dialogue is ordinarily determined by its text. This precludes the application of pre-existing formal schemes. As the pieces written during the first three decades of the seventeenth century rarely exceed the number of 150 bars, the lack of a musical framework is hardly felt as such. After 1630 dialogues gradually grew in length, running to 150–250 bars around the middle of

---

[23] The autographs of Daniel Bollius, formerly in the University Library of Wrocław (Breslau), were destroyed during the Second World War. However, a few fragments reproduced in 1931 survive (see H. J. Moser, *Die mehrstimmige Vertonung des Evangeliums* (Leipzig, 1931; repr. 1968), 54–9).

[24] Transcription in Part Two. For the authenticity of the viol parts, see the discussion of this work in Chapter 4.

the century. The resulting danger of musical incoherence induced composers to apply unifying devices with or without textual additions. In his first book of motets (1610) Alessandro Grandi had included an Annunciation dialogue of unusual length (229 bars), to the scriptural text of which he added the words 'Tota pulchra es Maria, et macula non est in te' (derived from a Vesper antiphon proper to the Marian feast of 8 December). These are repeatedly sung by two 'celestial voices', unifying the composition as a whole. True vocal ritornellos appear in dialogues by Cossoni and Sister Cozzolani.[25] Other composers went less far, merely repeating an ensemble section at the end (Chinelli, Carissimi).[26]

The division of dialogues and motets into sections occurred first in Rome but soon spread over the whole of Italy; this formal procedure was also adopted in northern and western Europe. The sections differed from each other first by scoring alone but later also by metre and prescribed tempo. Intervallic devices strengthening the structural coherence within the individual sections were sometimes applied with amazing subtlety. In the dialogue 'Domine, quis habitabit' by Carissimi, all the six questions start with descending fifths or fourths, while Legrenzi, in his 'Peccavi nimis in vita mea', opposes the beseeching sinner to a wrathful God by his use of ascending and descending sixths respectively, a procedure shown in Ex. 1.

Ex. 1

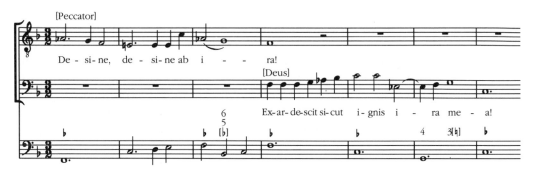

Occasional use of a *basso ostinato* in works by Giovanni Antonio Grossi and Benedictus a Sancto Josepho was restricted to the opening sections.[27] A piece consistently written in the form of strophic variations is 'Ecce spina, unde rosa' by the Sicilian Giuseppe Caruso; based on the 'aria della panaviglia', it must be considered a rare exception within the repertory.[28]

[25] C. D. Cossoni, 'O superi'; C. M. Cozzolani, 'Psallite superi'.
[26] G. B. Chinelli, 'Quid superbis caetus mortalium'; G. Carissimi, 'Domine, quis habitabit'.
[27] G. A. Grossi, 'O fortunati dies'; Benedictus a Sancto Josepho, 'Audite virgines'.
[28] Transcription in Part Two.

Sectional writing eventually led to the reduction in number of sections set in semi-recitative style and the predominance of arias. Originally, the term aria denoted nothing more than 'written in aria style'; hence in dialogues by Maurizio Cazzati one encounters 'arias' of only seven or nine bars.[29] However, around the turn of the century true arias in da capo form appear in works by Bassani, Ghezzi, and the Frenchman André Campra. The introduction of this formal type seriously undermined the character of the dialogue, assimilating the genre to that of the secular dramatic cantata.

Melodic Style

The appropriate means of representing sung speech, meeting both textual and musical requirements, was the semi-recitative. This vocal style allowed not only relatively free declamation but also florid passages that occurred mostly at the end of a phrase. The direction 'Si canta senza battuta' found in a dialogue by Antonio Cossandi (1640) gives us an idea of the manner of performance.[30] On the other hand, the semi-recitative often contains cantabile passages that are sung in regular tempo. Within a single phrase the technique of composition may change from declamation to pure melody as, for instance, in the opening section of Cossoni's 'Ave Crux', sung by the personified Santa Chiesa (Ex. 2).

An effective declamatory style depends largely on correct musical prosody; so it is hardly surprising that throughout the century composers took special care over the rendering of word-rhythm. Typical of this scrupulous attitude is a passage in the St Cecilia dialogue by Marc-Antoine Charpentier, 'Est secretum, Valeriane' (H 394).[31] In his setting of the words 'O suavis melodia' the composer first stressed the third syllable of 'melodia', placing it on a strong beat. Subsequently he must have realized that in post-classical Latin the second syllable should be stressed rather than the third (as in French and Italian). Leaving the music untouched he then changed the order of the words (Ex. 3).

A curious rhythmic formula is frequently encountered in dialogues by Giovanni Antonio Grossi (Ex. 4), Giuseppe Allevi (Ex. 5), and Isodoro Tortona (Ex. 6). This entails the division of a minim in common time into three units in the ratio $3:3:2$.[32] All three composers worked around the middle of the century in the city of Piacenza. Grossi was *maestro di cappella* at the Cathedral from 1644 to 1648, while Allevi was employed elsewhere in the city at the same time. After Grossi's departure Allevi became his successor, keeping the position until his death (1670). Tortona was Allevi's favourite pupil, as is shown by the inclu-

---

[29] M. Cazzati, 'Ad cantus, ad gaudia' and 'Silentium omnes' (both dated 1668).

[30] A. Cossandi, 'Cur praecepit vobis Deus'.

[31] The 'H' number refers to H. W. Hitchcock, *Les Œuvres de/The Works of Marc-Antoine Charpentier: Catalogue raisonné* (Paris, 1982).

[32] G. A. Grossi, 'Heu! Domine, respice et vide'; G. Allevi, 'Quam mihi obscura'; I. Tortona, 'Quid sentio?'

Ex. 2

sion of several dialogues and motets in his master's *Compositioni sacre*. Since the rhythmic design cannot be traced in the works of other composers, it was in all probability Grossi's invention. Subsequently Allevi must have adopted and in turn transmitted it to Tortona. So we may rightly speak of the 'Piacenza formula'.

Early in the century the semi-recitative, being almost always set in common time, was occasionally 'interrupted' by a few bars in triple time. Far from affecting its character, these changes of metre were the logical consequence of the declamatory setting of the words. When, however, after 1630 dialogues grew in length, the triple-time fragments developed into separate sections contrasting

Ex. 3

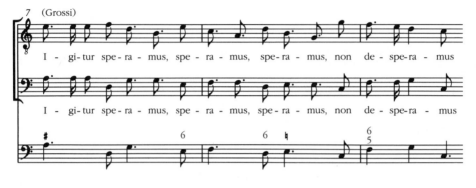

O su - a - vis me - lo - di - a! (deleted)
O me - lo - di - a su - a - vis!

Ex. 4

7 (Grossi)

I - gi-tur spe-ra - mus, spe - ra - mus, spe-ra - mus, non de - spe-ra - mus

I - gi-tur spe-ra - mus, spe - ra - mus, spe-ra - mus, non de - spe-ra - mus

Ex. 5

(Allevi)

in ri - - sus lan - guo - res con - ver - - te

Ex. 6

(Tortona)

Dic mi- hi, dic mi - hi, sum e - te-nim

with those set in semi-recitative style. While triple-time sections were generally used for the setting of joyful words, they could also serve the expression of sorrow or remorse (Ex. 7).[33]

Dialogues abound with rhetorical figures, illustrating the words and their emotional background. Among the most frequently applied are the *climax* or *auxesis* (the repetition of a melodic entity a second or third higher), the *exclamatio* (the upward leap of a minor sixth), the *pathopoeia* (a chromatic passage), the *saltus duriusculus* (a dissonant leap, often clashing with harmony), various

[33] G. A. Grossi, 'Heu me! miserum me!'

types of *hypotyposis* (word-painting), the *faux-bourdon* (homophonically descending sixth chords denoting sadness), and the *aposiopesis* (a general pause, mostly connected with death or eternity). The use of these figures reminds us of the dialogue's true character, which is that of a rhetorical, rather than a lyrical, composition. It is also for this reason that for a long time its melodic style resisted the growing tendency, apparent in other musical genres, to employ mechanical rhythms. When, eventually, rhetoric became subordinated to abstract melodic patterns, the dialogue faded from the musical scene.

Ex. 7

Melodic–Harmonic Writing and Counterpoint

Some aspects of the dialogue involve the harmony as well as the melody. This regards the influence of the Florentine style and the application of *seconda prattica* procedures. Severo Bonini, who spent the greater part of his life in Florence, was a pupil of Caccini; the Canticle dialogues included in his *Affetti spirituali* (Venice, 1615) are explicitly composed 'in stile recitativo di Firenze'. Yet the short-breathed melodic lines compare unfavourably with those of Caccini's solo-madrigals. The vocal writing is dull (even if one assumes addition of improvised embellishments) and depends too much on the underlying harmony. The dialogues of Francesco Capello show more affinity with the Florentine style.[34] The expressive melodies with occasional chromaticism, sometimes dependent on, sometimes independent from, the *fondamento*, mark him as one of the most progressive composers of sacred music during the first decades of the seventeenth century.[35] His unorthodox melo-harmonic devices are almost always related to the text and can be considered examples of the *seconda prattica*. Occasionally these are also found in the works of a few contemporary composers like Banchieri and Bazzino. In general, however, the dialogue's chordal support followed the harmonic conventions of the time. This is also true of pieces written after 1630. Instances of deviation from harmonic rules, encountered in works by Cossoni, Grossi (see Ex. 7 above), Della Ciaia, and Domenico Mazzocchi, are exceptional. The essence of the dialogue lies in the melody, rather than the harmony.

Outside Italy, it was Marc-Antoine Charpentier whose harmonies strongly reinforced the musical expression of the dialogued text. His frequent use of the augmented triad is particularly noteworthy. Daring procedures are also found in the works of Benedictus a Sancto Josepho, who showed a liking for the *cadentiae duriusculae*, that is, unusual dissonances in the penultimate chords of a cadence.

The dialogue proper had little use for counterpoint. Occasionally one comes across short imitations between vocal parts or melodic anticipation by the continuo. Contrapuntal handling of the musical material in the exchange of speech, as found in Grandi's dialogue 'Heu mihi!', is a rare exception.[36] Non-dialogued sections, in particular a ritornello or a conclusio, offered more possibilities. Alleluias, serving as a refrain or final statement, were sometimes written as a canon or a little fugue (examples include works by Giovanni Paolo Colonna and Gasparo Casati).[37] Besides, canonic setting of a phrase in the middle of a piece,

---

[34] See, for instance, 'Dic mihi, sacratissima Virgo'.
[35] See J. Kurtzmann, 'Giovanni Francesco Capello, an Avant-gardist of the Early Seventeenth Century', *Musica disciplina*, 31 (1977), 155 ff.
[36] Transcription in Part Two.
[37] G. P. Colonna, 'Adstabat coram sacro altari'; G. Casati, 'Quid vidisti, o Magi?'

like the fragment taken from a Nativity dialogue by Pietro Bertolini (Ex. 8), contributes to the variety of musical expression.[38]

Texts sung by ensembles representing the narrator or a large group of persons (*turba*) were usually set homophonically. However, this did not exclude a melodic treatment of the middle voices, resulting in a harmonically governed counterpoint. Roman composers such as Ratti, Quagliati, and the brothers Mazzocchi were particularly expert in this technique. Without sacrificing the intelligibility of the words, their eight-part ensembles offer much musical interest.

Ex. 8

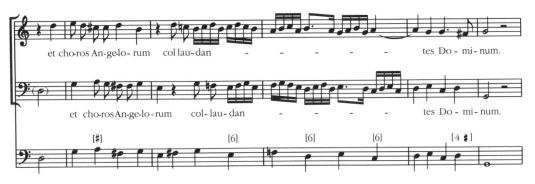

Dramatic Qualities

The use of the terms 'drama' and 'dramatic', originally related to visual action on the stage, becomes problematic if applied to purely aural representation. In particular, the meaning of the adjective 'dramatic' has been watered down in modern usage. It often denotes an effect rather than a quality, as, for example, in a statement like 'the dramatic entry of the horns shortly before the recapitulation'. Used in this way, the word has almost become a synonym of 'thrilling' or 'shocking'. When we are speaking of texted music, such as the dialogue, the term 'dramatic' will usually be employed in a sense closer to its original meaning, but even then it may easily lead to misunderstanding. Howard Smither was faced with this problem when, in his pioneering article from 1967, he divided the dialogues written during the first three decades of the seventeenth century into different categories:

All the texts may be considered dramatic in a general sense, since all contain verbal exchanges comparable to those found in a drama. In a more specific sense, however, the

---

[38] P. Bertolini, 'O felices pastores'.

texts have been classified . . . as 'dramatic', 'narrative–dramatic', and 'reflective'. In this sense the term 'dramatic' is applied to a text in which the action is revealed only through dialogue between characters without the aid of a narrator, and 'narrative–dramatic' are those using a narrator in addition to dialogue between characters. A 'reflective' text is a dialogue without a narrator, comparable to a reflective moment in a drama; no action or event is set forth, but the characters usually praise each other (as in the texts from the Song of Songs) or praise a saint, Jesus, etc. The reflective type differs from other motets with texts of praise only in that the separation of the roles, each sung by a different soloist, is strictly maintained.[39]

Smither's classification was undoubtedly useful for his purpose: the description and analysis of sixty-one dialogues composed during the period 1600–30. I see no reason to reject the chosen criteria, the more so as I have nothing better to offer. A few points should be discussed, however. (*a*) Does the term dramatic, used in a general sense, denote a quality common to all dialogues? (*b*) Is the division into three categories also applicable to the pieces written after 1630? (*c*) Are not narrative portions included in dialogued texts sometimes transformed in the mind of the listener into action? (*d*) Assuming that 'action' necessarily implies a situation at the end differing from that at the beginning, can it be maintained that in reflective dialogues 'no action or event is set forth'? Let us try to answer these questions.

(*a*) In the title of his anthology, *Sacred Dramatic Dialogues, 1600–1630,* Smither obviously interpreted the term in its general sense. It seems to me that, employed in this way, the adjective 'dramatic' has become a modifier that does not modify anything. If all dialogues are called 'dramatic', simply because they present a conversation, then the coupling of the two words amounts to a mere tautology. It should be pointed out, however, that non-conversational pieces, described earlier in this chapter, were nevertheless termed *dialogo* by their composers.

(*b*) It is true that the division into three types can also be applied to the works written after 1630. Yet there are a few texts which, although reflective, contain a narrative role, for instance 'Adstabat coram sacro altari' by Giovanni Paolo Colonna. Strictly speaking, this type, though admittedly rare, belongs to a fourth category, that of the narrative–reflective dialogue. In addition, one comes across examples of texts praising a saint which nevertheless include moments of action—for example, 'Quam mihi obscura' (1668), set by Giuseppe Allevi.

(*c*) The mental transformation of narration into drama on the part of the listener should not be overlooked. Some narrative text portions are so emotionally charged that they are easily experienced as 'dramatic'. The essential difference with the use of the term in the mere sense of 'thrilling' or 'shocking' is obvious. The narrator speaks of rationally conceivable events and feelings which, in prin-

---

[39] Smither, 'Latin Dramatic Dialogue', 408–9.

ciple, could also be represented on a stage. Moreover, he often makes his own emotional response to the related occurrences. In the realm of secular music one thinks of Monteverdi's *Combattimento di Tancredi e Clorinda*. No one would deny the dramatic qualities of this work; yet about 90 per cent of the text is sung by the narrator. Although sacred dialogues are ordinarily more sober in expression, it is easy to imagine a 'staged' performance of such scenes as those of the uncompassionate servant (Ratti)[40] or the battle between Heraclius and Chosroes (Cossoni).[41] Nevertheless the narrator plays an important part in both pieces.

(*d*) Smither classes the texts taken from the Canticles as reflective, since they lack action or events. This statement seems too categorical; it is contradicted by several dialogued episodes. For instance, the lines starting with the words 'Adjuro vos' (5: 18ff. and 6: 1–2) contain a coherent, though uncompleted, speech action. A female, apparently a young person, asks a group of women, addressed as 'daughters of Jerusalem', whether any of them has seen her beloved. The women ask her in turn to describe her lover's complexion. Then follows a eulogizing enumeration of the young man's physical qualities: his skin, his head and hair, his eyes, cheeks, lips, and so on (seven verses, most of which are usually omitted). Subsequently, the women enquire where he has gone, so that they can help to seek him. The girl tells them that her lover went into the garden to feed his flock and to gather lilies. Such is the scriptural text, set by quite a number of composers. However, Leone Leoni took a further step in his treatment of this subject. By adding a few freely invented words to the text of the group of women ('Ecce dilectus tuus!' (Behold your lover!)) he rounded off the episode, and so the concluding phrase—'I am my beloved's and my beloved is mine' (6: 2)—sung by all the voices, has become entirely convincing. Following Smither's criteria, this text belongs to the category of dramatic dialogues, rather than that of reflective dialogues, and the same is true of some other subjects derived from Canticle verses that are widely separated in the Bible but are now joined together in order to obtain a coherent discourse.

Despite its problematic aspects, the qualification 'dramatic', applied to a dialogue, will be used in the following chapters, albeit mostly in a specific, narrow sense. Truly dramatic dialogues are those which involve a conflict, occurring either between two or more characters, or in the mind of a single person. Pieces of this kind offer a great variety of subjects, including the scene of Mary Magdalene at the tomb of Christ, the sacrifice of Abraham, the battle between St Michael and Lucifer, a martyr resisting a tyrant, or a soul struggling with the devil. My restricted use of the term 'dramatic' implies that subjects like those of the Annunciation or the Nativity are excluded from this category. However moving these events may be, they cannot be called dramatic. As regards the musical

---

[40] L. Ratti, 'Simile est regnum caelorum'.   [41] C. D. Cossoni, 'Ave Crux'.

means employed for the expression and reinforcement of dramatic texts, they are too varied to be discussed in this chapter. Suffice it to mention the rapid exchange of fragmentary phrases or even single words by means of imitated or contrasted motifs, the use of the *stile concitato*, vocal as well as instrumental, and the alternation of slow and fast tempos in quick succession.

There is not the slightest indication that Latin dialogues were ever performed as theatrical pieces (Giovanni Battista Doni spoke of 'dialoghi fuor di scena').[42] Yet a few 'stage directions' occurring in works by Alessandro Grandi and Chiara Margarita Cozzolani suggest a 'mental' transformation of the church into a theatre. In his Annunciation dialogue, discussed above, Grandi indicated that the performers of the celestial melodies should be 'far away' and 'hidden'. As for Sister Cozzolani, she offers the listener the almost graphic image of the shepherds hastening to Bethlehem by means of the direction: 'All of them should gradually soften their voices as if they go away.'[43] A dialogue by Benedictus a Sancto Josepho is noteworthy for its inclusion of an 'aside', that is, three characters plotting against a fourth one, who is not supposed to overhear them.[44] These examples show us how close the dialogue could come to the world of the stage.

Northern Italy and Rome

The role dialogue originated in northern Italy, that is, the region embracing the Venetian Republic, the duchies of Milan, Parma, Mantua, and Modena, and the northern part of the Papal States. Curiously enough, the contribution to the repertory in Piedmont and the Genoese Republic was less than meagre. Only Giovanni Battista Fergusio, working in Turin, wrote a number of dialogues.

North Italian composers usually set their texts for two or three soloists with or without a concluding ensemble of four or five voices. This type soon spread over the whole peninsula and Sicily. Dialogues were already composed during the second decade of the century by musicians living in the extreme south, such as the Apulian Donato de Benedictis and the Sicilian Michele Malerba. Yet the bulk of the repertory came from Northern Italy. Here the chief masters of the genre were Alessandro Grandi and Ignazio Donati. Almost all of Grandi's dialogues date from the time he was employed in Ferrara as *maestro di cappella* of the Accademia dello Santo Spirito (1610–15) and the Cathedral (1615–17). Donati, one of the most travelled musicians of his time, worked successively in Urbino, Pesaro, Fano, Ferrara, Casalmaggiore, Novara, Lodi, and eventually Milan, where

---

[42] G. B. Doni, *Compendio del trattato de' generi e de' modi della musica* (Rome, 1635). The phrase is cited in David Nutter's article 'Dialogue', *New Grove*, v. 415.

[43] C. M. Cozzolani, 'Gloria in altissimis Deo' (1650). Transcription in Part Two.

[44] Benedictus a Sancto Josepho, 'Posita in medio' (1678).

he ended his career as *maestro di cappella* of the Cathedral. Both composers excelled in their melodic writing and counterpoint integrated in the *stile concertato*.

Surprisingly, during the greater part of the seventeenth century none of the *maestri di cappella* at St Mark's in Venice took any interest in the Latin dialogue. Neither Monteverdi nor his successors Rovetta and Cavalli contributed to this repertory. Milan, on the other hand, became a centre of dialogue composers, especially after 1630. This is the more remarkable since the Lombardic capital was known for its conservative taste in sacred music, resulting from the austere precepts imposed by the champion of the Counter-Reformation, Cardinal Carlo Borromeo. Moreover, the duchy was ruled by Spaniards who were anything but favourable to innovations. Yet during the period 1630–90 nearly all the *maestri di cappella* at the Cathedral (Donati, Crivelli, Grancino, G. A. Grossi, Cossoni) wrote dialogues in the modern *stile concertato*, and the same is true of composers working elsewhere in the city (Francesco Bagati, Giovanni Battista Beria, Chiara Margarita Cozzolani, Carlo Giuseppe Sanromano).

Another centre was Bologna. During the 1660s several members of its most prestigious musical establishment, the *cappella* of San Petronio, enriched the repertory: the *maestro* Maurizio Cazzati, the first organist Cossoni (most of whose dialogues were composed in this city), and the second organist Giovanni Paolo Colonna.

While it is true that Giovanni Antonio Grossi, the most prolific of all the composers discussed in this study, wrote a number of dialogues for services in Milan Cathedral, many other pieces must date from the years he spent in Piacenza and Novara. In Novara his predecessor at the Cathedral, Gasparo Casati, had already cultivated the genre during the 1630s.

Throughout the first three decades of the century scriptural dialogues dominated the north Italian repertory. Freely invented texts, like those of Banchieri's *Cuor contrito al suo Creatore* and Ricio's 'Cur plaudit hodie' (in praise of an unspecified saint) were rather exceptional. After 1630 the number of hagiographic and moralizing texts rose rapidly. The *conclusio* was sung by the protagonists alone; additional voices, like those found in the works of Donati and Bazzino, were no longer employed. With few exceptions, the *historicus* or *textus* also disappeared from the scene. Narration was avoided or, if indispensable, implied in the speech of the characters. Although the scriptural dialogue did not vanish from the repertory, the biblical words were paraphrased rather than followed strictly. An episode from the Old or the New Testament was sometimes taken as the starting-point for the development of a freely invented text containing a moral message—for instance, that of the Prodigal Son conversing with two allegorical characters, La Speranza and La Desperazione (Grossi). The battle between St Michael and Lucifer, based on a fragment of Revelation (12: 7–9) and elaborated with many invented details, became a particularly favoured subject in the middle of the century. Nativity dialogues, featuring one or more angels and the shepherds, were still frequently set to music. The Annunciation, on the other hand, practically disappeared from the repertory. Nearly all the subjects were rendered with the aid of only two or three soloists, more elaborate scoring becoming increasingly exceptional.

Whereas the northern style prevailed all over Italy during the whole century, one city produced a different type of dialogue: Rome. In the second decade Agostino Agazzari, Ottavio Catalani, and Abundio Antonelli had already set various texts based on scriptural subjects for eight voices, a scoring also used by Lorenzo Ratti and Paolo Quagliati during the 1620s. The works of these composers differ from those written in the north in that they assign a more important role to the 'chorus', representing the *historicus* or the *turba*. These ensembles are set for double choir. Yet the narrator also speaks by means of a single voice, a procedure which contributes to the variety of scoring within the composition. Other soloists are used for biblical words in direct speech, each of them keeping strictly to his role in the scene. The writing for single voices as well as ensembles shows a restrained expression contrasting with that of the more emotional north Italian works. Needless to say, any trace of the Florentine style and the *seconda prattica* is lacking. In the Roman eight-part dialogue late-Renaissance ensemble technique became fused with more modern stylistic traits.

After 1630 this type of dialogue maintained itself in pieces written for the church as well as the 'Crocifisso' oratory of San Marcello. Some composers, including Domenico Mazzocchi and Francesco Della Ciaia, intensified the musical expression, rendering emotional passages in the text by means of chromatic melodies and unusual harmonic procedures. The five extant oratorio dialogues by Marco Marazzoli, on the other hand, are austere in their expression, using almost exclusively recitative in the parts of the soloists. Written for the services on the Fridays during Lent, their style contrasts with that of the composer's Italian oratorios, which contain numerous ariosos and short arias.

The Sienese nobleman Alessandro Della Ciaia was one of the very few musicians living outside Rome who adopted the style of that city in their dialogues. As for northern Italy, only Leone Leoni, Giovanni Battista Fergusio, and Giovanni Battista Stefanini may be considered imitators to any extent. Stefanini added to the title of his fourth book of motets (including a dialogue in honour of St Elisabeth[45]) the words 'all'uso di Roma' (1618). In general, however, northern Italy remained unaffected by the Roman style. Only the sectionalization of both motets and dialogues, typical for Rome, was readily adopted.

The traditional image of Rome was that of a bastion of ultra-conservative taste, and for a long time this reputation remained unquestioned in musicological studies dealing with the early Baroque period. Obviously, the use of the *stile*

---

[45] G. B. Stefanini, 'Exurgens Maria'.

antico and polychoral scoring in settings of the Mass and ceremonial motet texts was considered characteristic of Roman church music in general. Only recently has this view been challenged, notably by Graham Dixon, who has shown that right from the beginning of the century composers wrote small-scale motets and dialogues in a style no less progressive than that of their north Italian colleagues.[46] They practised the small-scale concertato, probably under the influence of Viadana, who had already visited Rome before 1600. The local pioneer was Agostino Agazzari; his two-part and three-part motets with *bassus ad organum*, though published in 1606, were composed as early as 1603. Agazzari was also the author of the first treatise of continuo playing (1607), a book that became very influential throughout Italy.[47] Two small-scale dialogues are included in his *Sertum roseum* (1611);[48] other pieces appeared in his volumes of 1613 and 1625.[49] Several Romans, including Vincenzo Pace and Francesco Anerio, followed Agazzari's example. Among the composers of few-voiced dialogues are also those who created the more specifically Roman type of double-choir dialogue (Antonelli, Catalani, Ratti, and Quagliati). The style of all these small-scale works does not differ essentially from that of the north Italian compositions; the same is true of the non-oratorical dialogues written in Rome around the middle of the century. These include pieces of high quality by Giacomo Carissimi and Bonifazio Graziani.

After Carissimi's death (1674) Roman composers seem to have lost interest in the Latin dialogue. As for the rest of Italy, after 1670 only a few pieces are preserved in print (Monferrato, Urio, Bassani, Ghezzi) or in manuscript (Cossoni, Rusca). The scarcity of sources from the last three decades of the century points to the fact that as a genre the dialogue gradually became extinct in Italy. The musical reasons for this decline, in particular the adoption of schematized forms, have already been discussed. However, extra-musical factors also contributed to its disintegration. Diminishing interest in the setting of Latin words, together with a growing predilection for Italian texts, point to the 'secularization' of sacred music in general. The dialogue lost its place within the liturgy, which was reduced to its indispensable items. The small-scale motet likewise passed away, except for brilliant solo pieces written in cantata style. All these symptoms are characteristic of the 'easy-going quietism' that held sway in the Church, contrasting with the 'triumphal spirit' during the early phase of the Baroque era.[50] In point of fact, they announce the 'worldly' religious attitude prevailing in the new century.

[46] G. Dixon, 'Progressive Tendencies in the Roman Motet during the Early Seventeenth Century', *Acta musicologica*, 53 (1981), 105–19.
[47] A. Agazzari, *Del sonare sopra 'l basso con tutti li stromenti e dell'uso loro nel conserto* (Siena, 1607).
[48] A. Agazzari, *Sertum roseum ex plantis Hiericho*, Op. 14 (Venice, 1611).
[49] A. Agazzari, *Dialogici concentus*, Op. 16 (Venice, 1613); *Eucaristicum melos*, Op. 20 (Rome, 1625).
[50] See G. Dixon, *Carissimi* (Oxford, 1986), 1.

2

Italy: Dialogues Based on the Old Testament

Scriptural dialogues are for the most part based on conversational biblical texts. During the early decades of the seventeenth century especially, they reproduce the scriptural readings almost literally; later they show a growing tendency on the part of the authors to paraphrase the original words and to add details. In some cases it is difficult to decide whether a piece should be classified as a biblical dialogue with alterations and invented additions or as a free dialogue with scriptural quotations. Instances of this kind will be discussed either in this chapter or Chapter 3, or in Chapter 4.

The various subjects are arranged as far as possible in chronological order. Dialogues based on the Book of Psalms and the Canticles often form a patchwork, presenting in succession verses that are widely apart in the Bible. Therefore, works belonging to these sub-categories will be discussed separately.

Stories and Episodes

The Fall of Man, the first story in the Book of Genesis involving extensive conversation, was chosen only rarely as a subject for a Latin dialogue. Biagio Tomasi's 'Dum ambularet Dominus', set for six voices, dates from 1611. This composer, who worked as organist at the Cathedral of Comacchio near Ferrara, presented the scene (Gen. 3: 8–19) in almost strict accordance with the biblical text, although a few portions were omitted. Three voices represent God (B), Adam (T), and Eve (A) respectively; in addition words of narration are assigned to a soprano. The concluding section, containing the judgements passed on the serpent, Eve, and Adam, is sung by a six-part ensemble.

It is particularly interesting to compare Tomasi's setting of the summoning of Adam (Ex. 10) with that of Corfini's dialogued motet (1591) (Ex. 9), mentioned in Chapter 1. The two examples show clearly the drastic change of style that had occurred during the years around the turn of the century.

Tomasi's piece contains the instruction that the words of God should be sung by a hidden performer, 'placed above the other characters'. These, too, are sup-

Ex. 9

Ex. 10

posed to be hidden, in conformity with the biblical text. Therefore only the narrator must have been visible to the contemporary audience. Since both Adam and Eve abandon their roles in the polyphonic *conclusio*, this section was probably performed in the normal way.

Melody and harmony discreetly express the emotional implications of the text. The pre-tonal treatment of the key of F major allows the lowering of the seventh and third degrees in illustration respectively of the words 'timui' (I was afraid) and 'comedi' (I did eat) sung by Adam.

Tomasi's melodic style compares favourably with that of a similar composition by Antonio Cossandi (1640).[1] This minor composer set a large portion of the biblical text, as a result of which his dialogue is in three sections. The first contains the conversation between the serpent and Eve (Gen. 3: 1–5), written in an unimaginative recitative style which is maintained in the next section dealing with the summons of Adam and the verdict meted out to the serpent. Only

---

[1] A. Cossandi, 'Cur praecepit vobis Deus'.

towards the end of this section does the dry recitative yield to a long melisma (10 bars) on the word 'calcaneum' (heel); this is obviously supposed to depict the serpent's squirming. The final section, set for all four voices, is based on a free text ('Miserere nostri, Domine'). The only interesting aspect of this insignificant work is the composer's direction that the two monodic sections be sung 'without metrical beats'; this is probably an allusion to Caccini's and Gagliano's concept of *sprezzatura*. The concluding section, on the other hand, prescribes strict time-keeping ('Si canti con la battuta').

The slaying of Abel, the first act of violence described in the biblical story of mankind, provided material for three dialogues, by Giovanni Francesco Capello (1610),[2] Adriano Banchieri (1625),[3] and an anonymous composer (undated).[4] The first two works deal with God's interrogation of Cain; they are scored for tenor (or soprano) and bass. In an article published in 1977 Jeffrey Kurtzmann characterized the Brescian organist Capello as an avant-gardist of early seventeenth-century sacred music in general and in addition as a pioneer of the dialogue genre. A study of his setting of 'Ubi est Abel' confirms this view. The text presents a slightly altered version of Gen. 4: 9–10.

DOMINUS. Ubi est Abel, frater tuus?
CAIN. Nescio, Domine. Numquid custos frater mei sum ego?
DOMINUS. Ecce vox sanguinis fratris tui clamat ad me de terra.

On this scant material the composer constructed an impressive scene. The opening question is sung several times, beginning on the notes *d*, *g*, and *c'* respectively, and each time introduced by a distinctive melodic line in the instrumental bass (Ex. 11). The same symbolic formula precedes the concluding words of God, expressed by means of a powerful melody covering no less than two octaves (Ex. 12).

Capello's reputation is based mainly on his skilful application of *seconda prattica* devices, as exemplified by dialogues dealing with subjects borrowed from the New Testament; these will be discussed further on in this chapter. The fact that he was also able to write music of high quality without having recourse to these fashionable procedures testifies to his great skill.

Banchieri's setting includes, apart from the interrogation, verses 11–15, containing God's verdict, Cain's admission of his guilt, the expression of his fear of being killed as an outlaw, and God's reassurance on this point. These details weaken the dramatic impact of the dialogue as a whole, the more so as Banchieri's musical rendering of the interrogation itself is far less impressive than that of Capello. The two-part setting of God's final words—a procedure already abandoned by most of the other composers during the 1620s—is particularly unsatisfactory.

---

[2] G. F. Capello, 'Ubi est Abel, frater tuus'.   [3] A. Banchieri, 'O Cain, ubi est frater tuus'.
[4] 'Offerebat Cain', incorrectly attributed to Carissimi by L. Bianchi.

Ex. 11

Ex. 12

The anonymous 'Offerebat Cain' differs from the two other works in that it presents the biblical text almost completely (Gen. 4: 3–5, 8–15). The piece is set for six voices, including the roles of Cain (A) and God (B). The narrative words are assigned first to a tenor, and then to three sopranos. It is particularly this ensemble of high voices that expresses more than the mere events. Cain's anger at the failure of his offering and his rancour towards his brother are effectively depicted by imitatively treated melismas on the words 'iratus' and 'consurrexit'. The interrogation is soberly rendered; like Capello, the composer concludes the verdict with a melody covering the entire compass of the bass. The final sextet, set to the words 'And Cain went out from the presence of the Lord and lived like an exile on earth', is a rather abstract polyphonic section; yet it makes a suitable completion of the scene.

The sacrifice of Abraham, undoubtedly one of the most dramatic episodes in the Book of Genesis, served as a subject for a fairly large number of dialogues. Six examples of these will be discussed in detail in Chapter 6. However, there also exists a work dealing with the announcement of the birth of Isaac as described in Gen. 18.[5] Its composer, Giovanni Antonio Grossi, has already been mentioned more than once in the previous chapter; in the course of this book we shall encounter him on many further occasions. A study of his enormous output (more than five hundred sacred works) is still lacking. Grossi was not only a prolific but also a particularly versatile musician, showing proficiency in all the styles and techniques of his time. Mariangela Doni's observation that 'his style is unoriginal'[6] may well be contested; it is contradicted, for instance by Ex. 7, quoted in Chapter 1. Also indicative of Grossi's imposing musicianship is the fact that in 1669 he won the *concorso* for the prestigious post of *maestro di cappella* at Milan Cathedral, being eventually preferred to his main rival, Giovanni Legrenzi.

'Heu! Domine, respice et vide', whose 'cast' features Abraham (B), Sarah (T), and three angels representing God (SSA), is typical of the expansion of the dialogue during the second half of the century; the piece runs to 313 bars. The announcement of Isaac's birth is preceded by a section paraphrasing Gen. 15: 2–3, that is, Abraham's complaint at having no offspring and a servant becoming his heir. The layout of the work is as follows:

	Bars	Text	Source	Style	Textual content
I	1–22	Heu Domine (B)	15: 3	semi-rec.	Abraham's complaint
II	23–37	Spera, nec despera (T)	free	semi-rec.	Sarah comforts her husband
III	38–50	Igitur speramus (TB)	free	duet	Expression of confidence in God
IV	51–69	Ecce video (B)	18: 2–5	recitative	Arrival of the angels; Abraham tells them that provisions will be prepared
V	70–99	Fac. Ecce adsumus (SSA)	18: 5	terzet	Angels: 'So do as thou hast said'

[5] G. A. Grossi, 'Heu! Domine, respice et vide'. Transcription in Part Two.
[6] M. Doni, 'Grossi, G. A.', *New Grove*, vii. 743.

	Bars	Text	Source	Style	Textual content
VI	100–6	Ecce parata sunt omnia (B)	free	semi-rec.	Invitation to the meal
VII	106–28	Edamus ergo (SSA)	free; 18: 9	terzet	Acceptance; question: 'Where is Sarah?'
VIII	129–39	Ecce in tabernaculo (B)	18: 9	semi-rec.	Abraham's answer: 'In the tent'
IX	139–56	Revertens veniam (SSA)	18: 10	terzet	Announcement of the forthcoming birth of a son
X	157–64	Oh, Deus meus (T)	18: 12	semi-rec.	Disbelief of Sarah; she laughs
XI	165–220	Quare risit Sara? (SSAT)	18: 13–15	quartet	Sarah rebuked; her denial and final admission
XII	220–43	Credite igitur (SSA)	free	terzet	Confirmation of God's promise to Abraham (see Gen. 13: 15–16 and 15: 5)
XIII	243–58	Credimus (TB)	free	duet	Expression of belief and gratitude
XIV	259–75	Ecce, quia credisti (SSA)	free	terzet	Angels: 'Thou shalt call the name of him Isaac'
XV	275–80	Gratias agimus (TB)	free	duet	Renewed expression of gratitude
XVI	281–313	Gratias agimus (SSATB)	free	quintet	Conclusio: Elaboration of the text of section XV.

The diversity of the formal components is matched by the inner contrasts of style. Melody, rhythm, and harmony evince highly original traits. The 'Piacenza formula' (see Chapter 1) appears no fewer than seventeen times in various configurations; in addition, other unusual rhythmic designs are employed, such as in bars 21 and 35–6. As for harmony, it appears that Grossi was fond of simultaneous cross-relations (see bars 27 and 71). Striking dissonances also appear in the *conclusio*, which shows prevalence of (contrapuntal) melody over harmony. The delightful imitation of Sarah's laugh in the parts of the angels (section XI) is effected by means of slow trills that should be sung *non legato* (bars 169 and 180–4). Taken as a whole, this dialogue is a model of its kind, rendering the scene with brilliance and lacking any hint of monotony.

Episodes from the story of Joseph and his brothers were set to music by three composers: Leandro Gallerano, Domenico Mazzocchi, and Marco Marazzoli. Working as an organist in Brescia and Padua, Gallerano was a typical provincial musician. His dialogue 'Audite fratres' (1640) deals with Joseph's first dreams and the fratricidal plot of his brothers: 'Behold, this dreamer cometh. Come therefore, and let us slay him . . . and we shall see what will become of his dreams' (Gen. 37: 6–10, 13–20). The following events (Reuben's intervention, the selling of the boy to the Ishmaelites, and the father's lament on seeing the blood-stained coat of his most beloved son) are all missing. Because of this, the work gives the impression of an unfinished story. Musically, the dialogue is quite undistinguished. Two of the four parts are assigned to Joseph (S) and Jacob (B), but no differentiation is made between the brothers and the polyphonically treated role of the narrator. The only noteworthy detail is the use of the rhetorical device of *circulatio* (a melodic design turning around its initial note), which in this case depicts the circle of sheaves in Joseph's dream.

Mazzocchi's *Dialogo di Gioseppe*, an oratorio dialogue, is quite another matter. Written for nine voices, it treats the complete episode, ending with Jacob's complaint. The brothers mostly sing in double choir, the second tenor and bass of which are also used for the roles of the narrator and Jacob. However, Joseph has a *canto solo* of his own. The work illustrates the composer's famed expressive handling of the semi-recitative, occasionally supported by unusual chords, and his elegant treatment of eight-part ensembles. The final tutti-restatement of the father's words with its striking dissonances is particularly moving.

The subject of Marazzoli's 'Erat fames in terra Canaan' is the visit of the brothers to Egypt (Gen. 42). This is likewise an oratorio dialogue, written for the SS. Crocifisso. Two violins are added to the five voices. Although this work, too, contains skilfully conceived ensembles, the recitative treatment of the solo parts is particularly austere, if not dry. There is no trace of the sensuous melodic writing, characteristic of Marazzoli's works set to Italian texts, such as his operas, cantatas, and *oratori volgari*.

It is a curious fact that many stories and episodes from the Old Testament that have served for full-scale oratorios are practically absent from the dialogue repertory in Italy. So we skip the books of Exodus, Leviticus, Numbers, Deuteronomy, Joshua, Judges, and Ruth and arrive at the books of Samuel and the Kings. Although I have not found an example of a piece treating the battle between David and Goliath, the message of the death of Saul and Jonathan served as a subject for an oratorio dialogue by Domenico Mazzocchi: *Lamento di David* (2 Sam. 1 ff.). The narrative portions of the scriptural text as well as the messenger's interrogation by David are set as recitatives. The killing of the messenger as recorded in 1: 15–16—embarrassing to a modern mind—is omitted. David's last question leading to this action ('How wast thou not afraid to stretch forth thine hand to destroy the Lord's anointed?') is retained, however. Since this is immediately followed by the lament, there is an awkward moment of discontinuation in the scene. In the lament itself, echoed and commented upon by two four-part ensembles, one of men, the other of women, Mazzocchi displays his expressive style to its greatest advantage. The women's choir, in particular, is full of striking dissonances that result from bold suspensions rather than *seconda prattica* devices.

Two dialogues dealing with the Judgement of Solomon present extremes of quality. Antonio Cossandi's 'Obsecro mi, Domine' (1640) is a particularly undistinguished piece. No effort is made to differentiate musically between the parts of the two women (both S) nor between them and the king (B). Only the words, set as they are in dry recitative, convey the conflict. The concluding section a 3, which endlessly repeats Solomon's last sentence ('Take what is yours and go in peace'), is written in unimaginative counterpoint.

To compare this rather clumsy work with Carissimi's oratorio dialogue *Judicium Salomonis* is almost grotesque. The Roman composer set his text for four voices with two violins. Although the final ensemble containing a praise for God is far from negligible, the most poignant passages are found in the section in which the two women confront each other. The setting of their heated argument is highly dramatic and, as has been pointed out by Graham Dixon, full of abrupt endings and interjections. The same author drew our attention to the meaningful opposition of the minor and major modes as well as consonance and dissonance in the women's respective reactions to Solomon's decision to cut the living child in half.[7] The whole work evinces a tendency to characterize each of the interlocutors in a way which goes far beyond the mere relation of the event.

A scene from the Book of Tobit was set to music by both Francesco Foggia and Carlo Donato Cossoni. This relates the expression of gratitude by the father for having been healed of blindness, and similar feelings of the son for having obtained a spouse. The benefactor, previously called Azarias, then makes himself known as the angel Raphael and refuses any reward. The concluding section contains a praise of God's mercy. All this is described in Tobit 12: 1–6 and 15.

A comparison of the two compositions clearly reveals the difference between the Roman and north Italian styles. Foggia's piece, an oratorio dialogue preserved in a Bologna manuscript, is set for five parts, including the roles of the father (B), the son (T), and the angel (S). The narrative words are invariably sung as short duets (AT) that form a welcome contrast to the recitative-like fragments of the soloists. The final quintet demonstrates expert homophonic and polyphonic writing. The sober character of this work, with its restrained expression of a non-dramatic text, may be considered typical of a Roman composer.

Cossoni, on the other hand, tried to compensate for the lack of dramatic tension by setting large portions of the text in aria style. This was possible because his dialogue paraphrases the biblical words. In the centre of the work there is a

---

[7] Dixon, *Carissimi*, 42, 45.

brief aria for the son, followed by an even shorter one for the angel (sixteen and eight bars respectively). Then the tempo changes from allegro to presto, and father and son sing a duet in three varied strophes, first each solo and finally together. The music is set syllabically to (rather lame) verses. Although Raphael's disclosure of his celestial identity opens as a semi-recitative, it ends in an aria-like style. The concluding terzet—there is no narrator—is set with effective alternation of the three voices. While it is true that the story is more or less conveyed to the listener, the text seems to have been written with the purpose of achieving an attractive musical setting rather than that of transmitting a scriptural lesson.

Among the numerous dialogues by Giovanni Antonio Grossi there are two pieces based on the books of Judith and Esther respectively. The story of Judith is enacted in the city of Bethunia alone, and so there is no part for Holofernes. The work is in three sections: Judith's announcement of her visit to the enemy camp, the expression of the people's sentiments hovering between hope and fear, voiced by the Dux (Ozias), and finally the triumphant return of the heroine. The textual construction of the scene is rather unsatisfactory. Judith speaks only in the opening section (bars 1–29), her return being reported through narration. No wonder that this work is not among the composer's best dialogues. Only the parts for the two narrators offer some interest. Far from merely relating the facts, the latter show their emotional involvement in the action.

The Esther dialogue, set for three voices, offered more possibilities. As in the piece about Judith, most of the text is a paraphrase of the biblical words. In at least one instance this proves an advantage. Instead of being informed of the imminent disaster by means of a go-between (the eunuch Hatach), the queen speaks in person with her pseudo-father Mordecai. While this enhances the dramatic intensity, the fact that Esther's revelation of Haman's plot to King Ahasuerus follows without any interruption makes an awkward impression. It is essential that Mordecai, demonstratively dressed in sackcloth, should sit *before* the gate, whereas the king should, of course, be *inside* the palace. The presentation of non-consecutive events posed a special problem to those composers who had no narrator at their disposal. In this case the solution adopted—that is, simply skipping the intervening time—proves unsatisfactory.

From a musical point of view the work offers considerable interest, showing Grossi's skill in both monodic and polyphonic writing. Esther's description to the king of the Jewish people's imminent fate is particularly moving (Ex. 13). In the *conclusio* the interlocutors retain their roles. All three rejoice in the execution of Haman, expressing themselves in words that seem neither educational nor edifying. This, however, does not prevent the composer from writing an attractive ensemble in transparent counterpoint.

Two works, one by Carissimi, the other by Grossi, deal with the biblical figure of Job. They demonstrate clearly the fact that it is virtually impossible to distinguish, on textual as well as musical grounds, between compositions written for

Ex. 13

services in an oratory and those intended for performance in church. Both pieces, entitled *Historia di Job/Giobbe*, present a practically identical dispute between an angel and the devil, leaving the sufferer in the middle until he chooses the side of the angel. In Carissimi's dialogue this event occurs soon after the start, but in Grossi's only after ninety-four bars. However, there is no question here of an authentic biblical story. As stated in Chapter 1, nowhere in the Scriptures does the devil converse with Job; moreover, there is no mention of an angel. While it is true that both works contain several biblical quotations, notably the cardinal words 'Sit nomen Domini benedictum' (Job 1: 21), they present a purely invented scene.

Carissimi's dialogue, conforming to the spirit of Lenten devotion, was undoubtedly written for a gathering in the S. Marcello oratory of the SS. Crocifisso. Grossi's composition, on the other hand, must have been intended for a regular service in the Cathedral either of Piacenza, Novara, or (most probably) Milan. Apart from their textual affinities, the musical resources of the two pieces show striking similarities. Both are scored for three vocal parts, that is, the angel (S), Job (A in Carissimi, T in Grossi), and the devil (B). Moreover, they are set in the same key (D minor) and have a total length of about 260 bars.

In other respects, however, a comparison of the two compositions evinces conspicuous contrasts. These have nothing to do with the functional difference between oratorical and church dialogues; they are accounted for by other reasons. One of these is the chronological separation. Although both works are preserved only in undated manuscripts, we may assume on stylistic grounds that Carissimi's dialogue was written before 1650; Grossi's style, on the other hand, points to the late 1660s or even the 1670s. This explains the great variety of musical procedures within his composition; the work abounds with contrasts of melody, rhythm, harmony, and tempo—features that are absent from Carissimi's piece. Then Grossi's dialogue contains ingenious instances of word-painting which go far beyond 'madrigalisms'. Job's description of his passively endured sufferings is depicted in the following passage (Ex. 14).

Ex. 14

Another reason that explains the compositional contrasts between the two dialogues is again the essential difference between the north Italian and Roman styles. Carissimi set his text predominantly in a semi-recitative style, lacking the spectacular traits found in Grossi's piece. Also typical of the Roman composer is the tight construction. The repeated words 'Sit nomen Domini benedictum' function as a refrain, occurring four times in the course of the composition. Written in a style approaching that of the aria, this refrain also forms a mild contrast with the intermediate solo sections. The text of Grossi's dialogue did not allow the application of this formal device, as it presents Job as a man who complains bitterly about his fate. It is only after a passionate plea by the angel that he finally praises God with a *cantabile* melody in triple time; this melodic style is cunningly adopted by the devil—without success, however. The biblical quotation ('Sit nomen . . .') appears in the final, climactic terzet; the composer's problem of how to fit the devil's role to these words was discussed in Chapter 1. Carissimi's dialogue also concludes with a terzet. However, since the character of his work is much more sober and intimate, the Roman composer could afford to act logically: his devil simply drops out, allowing the final restatement of the refrain to be sung as a duet.

An early example of the special way in which some Roman composers adopted the external character of north Italian small-scale motets and dialogues is the setting of a fragment from the Book of Isaiah by Domenico Massenzio:

'Quis est iste qui venit de Edom' (Isa. 63: 1–3, 7). Written for high voices (SSA), the piece shows a particularly restrained musical expression of the text, combining modal traits with modern cadential devices. Another piece based on a subject from this scriptural book is Carissimi's *Historia di Ezechia* (Isa. 38: 1–8). This oratorio dialogue takes up a theme introduced into the Lenten liturgy by the lesson from the Old Testament read in place of the Epistle at Mass on the day after Ash Wednesday. It was almost certainly written later than the *Historia di Job*. It tells the story of Hezekiah, king of Judah, to whom Isaiah, directed by God, announced the news of his imminent death. After the king's prayer that the divine decision be rescinded, he is granted fifteen further years of life. The work focuses on the prayer, as well as the final thanksgiving and praise of God, elaborated in a five-part ensemble. The solo sections show a melodic–harmonic style which can be described as expressively restrained. The role of the two violins in the prayer is particularly interesting. They repeatedly play a short ritornello, whose chromatic bass complements the words (Ex. 15). The addition of violins to the voices in the *conclusio* is scripturally justified in a literal sense by the verse 38: 20, which reads: 'The Lord was ready to save me; therefore we will sing my songs to the stringed instruments all the days of our life in the house of the Lord.'

Ex. 15

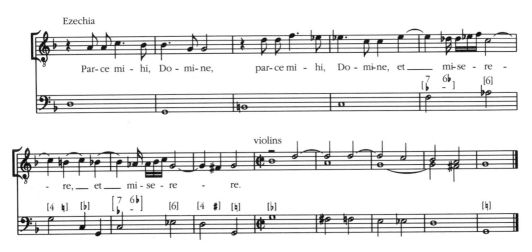

A survey of the church dialogues based on stories and episodes from the Old Testament raises this question: why was it that so many attractive and dramatically arresting subjects, such as the stories of Noah, Esau and Jacob, Moses, Joshua, Samson, Saul, David and Goliath, etc., were neglected? It could be that the choice of a restricted number of topics resulted from considerations relating to the matter of religious education. Another explanation is that among the missing subjects several call for a musical scoring that could hardly be realized with the modest means available to the average Italian church. Yet these answers are not entirely convincing. Besides, the number of works available for this study does not exhaust the repertory. For instance, I was unable to gain access to an anonymous dialogue between Moses and the Egyptian Pharaoh, written for two basses.[8] So the question must be left open.

Psalm and Canticle Dialogues

In the Book of Psalms exchange of direct speech is extremely rare. In point of fact, this occurs only in Psalm 14 ('Domine, quis habitabit'), and so it is hardly astonishing that this text was frequently used for dialogued compositions. It reads in English as follows:

1. Lord, who shall abide in thy tabernacle? Who shall dwell in thy holy hill?
2. He that walketh uprightly and worketh righteousness, and speaketh the truth in his heart.
3. He that backbiteth not with his tongue, nor does evil to his neighbour, nor taketh up a reproach against his neighbour.
4. In whose eyes a vile person is contemned; but he honoureth them that fear the Lord. He that sweareth to his own hurt, and changes not.
5. He that putteth not out his money to usury, nor taketh reward against the innocent. He that does these things shall never be moved.

Among the composers who used this psalm for a dialogue are Vincenzo de Grandis (Il Romano), Carlo Milanuzzi, Nicolò Stamegna, Stefano Landi, Vincenzo Ugolini, and Giacomo Carissimi, not to mention many others. Since the original text contains only a single question (twice formulated with different words), which receives a complex answer, these composers were faced with the problem of how to turn the words into a true conversation. Milanuzzi offers the most simple solution. The opening question as expressed in verse 1 occurs three times in the course of his composition and is answered by the words of verses 2, 3, and 5 respectively. Both Ugolini and Carissimi present verse 1 as two separate questions, each of which receives an answer. In addition, the texts of their dialogues contain other questions that are simply variants of the first one. In all three works the (human) questioner is represented by ensembles (AT in Milanuzzi, SS in Carissimi, and SAB in Ugolini); on the other hand, the respondent (presumably God) sings solo (T in Ugolini and Carissimi, B in Milanuzzi). In view of the undramatic words, the quality of these works depends largely on the purely musical capacities of the respective composers. Hence a skilful but very conservative musician like Ugolini could afford to treat the text in a stately, unemo-

---

[8] *Dialogue entre Moyse et Pharaon ou la sortie d'Israel d'Egypte.*

tional way. The style of Milanuzzi's piece is more lively and modern; because of the repeats of the opening words—musically only slightly varied—his composition is formally more satisfying. Carissimi's setting is a small masterpiece. The descending fifths and fourths unifying the questions have already been mentioned in Chapter 1. The work culminates, however, in a final section set to a free text: 'O sedes amata, a domus beata, ubi in Domino gaudebunt et exsultabunt sancti in conspectu Dei.' This opens with a homophonic passage followed by a splendid contrapuntal elaboration of a canzona-like theme (Ex. 16). This *conclusio* is then repeated, starting on a higher pitch. The contrapuntal segment now includes a fugal stretto; in the final bars the composer reverts to homophony.

Some of the works based on Psalm 14 also contain quotations from other psalms. To write a dialogue consisting of various psalm fragments is only a small step further. This is the case of the anonymous 'Vias tuas, Domine', set for two parts (SB). However, the composer of this piece also resorted to the Canticles, obviously with the intention of lending more animation to the static psalm verses. Consequently the two characters (Sponsa and Sponsus) express love of God rather than love for each other.

The transformation of a narrated conversation into direct speech was brilliantly realized by Bonifazio Graziani.[9] This is based on the famous Psalm 136, which deals with the exiled Israelites in Babylon. Since Graziani set the text for four high voices, the piece might have been composed for a service in a convent. Two pairs of voices (SS and SA) represent the Israelites and the Babylonians respectively. Following the paraphrased opening verse of the psalm ('Babylonis ad flumina'), assigned to the first and second sopranos, the Babylonians ask their captives to sing the hymn of Jerusalem. The Israelites' reaction is emotionally expressed: 'Heu, quomodo cantabimus dulcia Sion carmina in terra alienigena?' Then a conversation evolves in which the Babylonians are insistent but meet with a refusal. This part of the dialogue consists of short arias as well as duets, expressing emotion on one side and incomprehension on the other. Finally, in an antiphonal quartet, the Israelites decide to sing. However, the nature of their song does not meet the wishes of their oppressors, since the concluding section is set to words expressing intense emotional suffering: 'Sit vox turbae dolentis intermixta suspiriis et lamentis.' Graziani was considered, after Carissimi, the greatest church composer of Rome. The present work certainly testifies to the high quality of his music.

Finally, there exists a setting of a Vesper psalm transformed into a dialogue by Chiara Margarita Cozzolani. Among the female composers in Italy—more numerous than in other countries—she seems to have been the only one to cultivate the dialogue genre (with the exception of Isabella Leonarda, who wrote at

---

[9] B. Graziani, 'Babylonis ad flumina'.

Ex. 16

least one dialogue: 'Sic ergo anima'). Cozzolani took her vows at the Milanese convent of Santa Radegonda in 1620 and published four collections of sacred music from 1640 onwards.

Psalm 111, 'Beatus vir qui timet dominum' (1650), is set 'a 8 voci concertati in forma di dialogo'. The procedure of transforming the biblical text into a true conversation is quite simple and rather naïve: a phrase or sub-phrase is sung as a question by a group of voices and subsequently confirmed by another ensemble. The treatment of the opening verse offers a characteristic example (Ex. 17). At the end of each section all eight voices emphatically repeat the words 'Beatus vir' (verse 1: bars 22–5, 39–42, 97–100, 130–4) or 'Iucundus homo' (verse 5: bars 161–73, 195–206, 226–37); finally the two statements are combined (bars 266–71). Although these words are in some instances identically set, they function as textual rather than musical refrains. No questions are asked in the doxology, which serves as *conclusio*. Obviously, any expression of ignorance or doubt concerning these doctrinal words was considered out of place.

Taken as a whole, the work shows a particularly virtuosic handling of different blocs of voices, which, however, are not consistently used for fixed 'roles'. Voices representing the questioner elsewhere serve the respondent, thus enhancing the variety within the piece. As has been said in Chapter 1, this curious composition cannot possibly have been intended for a regular Vesper service in a parish church; it was probably written for a monastery or another ecclesiastical institution.

Like the Book of Psalms, that of the Song of Solomon is classed in the Vulgate among the *libri didactici*. Yet these verses, which by virtue of their literary value belong to the most celebrated in the Scriptures, possess lyrical rather than didactic qualities. Nor are they, generally speaking, narrative or dramatic. This does not exclude the presence of a few 'scenes'; these, however, being more or less fragmentary, cannot be called rounded episodes. As in the psalms, exchange of direct speech occurs only rarely, but phrases within the monologues quite often receive answers in quoted speech. To turn these verses into true dialogue must have been easy. A higher degree of proficiency was required from the authors if they were to give coherence to the inconsecutive and fragmentary texts; this was achieved by patching together verses that are widely separated in the Bible, as well as by inserting freely invented words.

No other scriptural book seems to have enjoyed more popularity among seventeenth-century composers of sacred music. This is apparent in both the motet and the dialogue repertories. It was undoubtedly the erotic sentiments expressed in beautiful language which attracted musicians within and beyond Italy. The way in which the Church reacted to their predilection for this book will be discussed below.

Only a limited number of dialogues based on the Song of Solomon will be examined here. Most are set to verses from the second and fifth chapters, occasionally combined with words found elsewhere in the book. Three 'scenes' can be discerned. Two of these, 'Ego dormio . . . Aperi mihi' (5: 2–6) and 'Adjuro vos' (5: 8–17 and 6: 1–2), are dramatic in as much as they collectively present a conflict that is eventually resolved. Despite their interrelation, these two scenes are always treated separately by composers. The third coherent fragment is of a purely lyrical nature: 'En dilectus meus loquitur mihi: Surge, propera, amica mea' (2: 10 ff.).

Among the earliest settings of this last scene (in which the Sponsus arouses his beloved) are dialogues by Ottavio Catalani and Alessandro Grandi, both dated 1616. Catalani, one of Carissimi's predecessors at the Collegium

Ex. 17

Germanicum in Rome, set the text a 2 with frequent melismas in both parts, a procedure befitting the lyrical words. Not surprisingly, Grandi's piece is of superior quality. Although it contains nothing spectacular, the continuous alternation of common and triple time (which, according to Jerome Roche, reflects the ebb and flow of amorous feeling)[10] confers on the composition a particular charm. The same words were set by Lodovico Bellanda, but for paired voices (ST and AB). Despite a few distant motivic relationships unifying the piece, this composition is quite unremarkable.

Lorenzo Ratti devoted an entire volume to the setting of verses from the Song of Songs, in the form of both motets and dialogues. The pieces in this collection are written in a north Italian style, completely different from that of his double-choir dialogues (to be discussed in Chapter 3). The text of his 'En dilectus meus loquitur mihi', though explicitly termed 'dialogo', is shared from bar 20 onwards by both the soprano and the bass. Therefore this composition must be considered a motet-like monologue for two voices rather than a true dialogue.

10 J. Roche, *North Italian Church Music in the Age of Monteverdi* (Oxford, 1984), 63–4.

The scene of the lover calling at the house of his beloved ('Open to me, my sister, my love') has a negative outcome. The Sponsus becomes irritated by the hesitation of the Sponsa and, when she finally opens the door, she finds that he has gone away. Then she seeks him everywhere but without result (5: 6). The setting of this text by the Sicilian Antonio Ferraro contains several instances of word-painting, such as 'umbrae' set in black notation and the general pause following the words 'Vocavi et non respondit mihi'. In other respects, however, the piece is quite mediocre. The same subject was treated by Leandro Gallerano.[11] Although by omitting the verses that describe the lover's disappearance he deprived the scene of its dramatic character, his composition is of a higher quality. Among its remarkable traits are the rapid exchange of scraps of phrases between Sponsus and Sponsa, the frequent occurrence of dissonant suspensions (ninths), and the chromatic setting of the words 'quia amore langueo' (Ex. 18).

Ex. 18

(Sponsus in unison with the b.c)

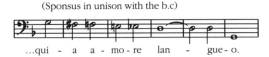

...qui - a   a - mo - re   lan   -   gue- o.

A work first attributed to a great composer and subsequently considered apocryphal meets with a bitter fate: no one takes interest in it any longer. This is the case with 'Tolle sponsa, tolle fores', a dialogue included in error by Lino Bianchi in the edition of Carissimi's complete works. Nevertheless, despite the fact that the authorship of Carissimi is out of the question, the piece makes a remarkable contribution to the repertory of Canticle dialogues. Its text differs from the previously discussed works in that it paraphrases the biblical words with very few literal quotations. Consequently, the author was able to present the 'story' in a more coherent form. As in the other pieces, the lover wakes his beloved by knocking on her door. Although recognizing his voice, she is too overcome by sleep to let him in. The indignant lover tells her that 'true love does not sleep'. When eventually she decides to get up, the Sponsus bitterly exclaims: 'No, no, go to sleep, you lazy one.' This part of the scene is enacted in a long dramatic duet, at the end of which the lover leaves. The girl's ensuing complaint is continued in the concluding section, sung by both voices. The melodic style of this anonymous dialogue, balancing between semi-recitative and aria, points to the middle of the century. Several details are noteworthy—for example, the setting of the words 'heu miseram, sopor obnubilat' (Ex. 19). From a musical point of view the work reaches a climax in the concluding section with its poignant chromaticism (Ex. 20).

---

[11] L. Gallerano, 'O pulcherrima mulierum'.

Ex. 19

The sequel to this scene, the girl's search for her lover, was a favourite subject for many composers. Nearly all begin their dialogues with the words 'Adjuro vos' (5: 8). This is the conversation between the Sponsa and the Daughters of Jerusalem, described in Chapter 1. The scoring is mostly for a solo soprano counterposed to a group of other voices; exceptions include the settings by Leone Leoni (for two four-part ensembles) and Abundio Antonelli (for soprano and bass only). The latter piece is remarkable because of the ingenious way in which the description of the lover's physical qualities is converted into true dialogue. The words

> Caput eius aurum optimum,
> comae eius sicut elatae palmarum

are presented as an exchange of short questions and answers (Ex. 21).

Ex. 21

Apart from these three scenes, other dialogues based on the Song of Solomon offer nothing more than a patchwork of individual verses taken from various chapters of the biblical book. These pieces present alternating expressions of erotic love in poetic language rather than coherent action. Severo Bonini's dialogues 'Tota pulchra es' and 'Vox dilecti mei' both belong to this sub-category. His compositions offer only meagre interest, except for the fact that they are explicitly written in the 'stile recitativo di Firenze'—an indication which, according to the preface to the collection, refers not so much to the compositional technique as to the performance practice. To call Bonini's dialogues 'embryonic

---

[12] M. Bonino, 'Bonini, S.', *New Grove*, iii. 21–2.

cantatas' seems rather far-fetched.[12]

Ratti's 'Fasciculus mirrhe' and, in particular, 'Sicut lilium' (both 1632) offer much more attractive music, especially in their concluding sections. Unlike 'En dilectus', mentioned above, these are true dialogues.

'Quam suave es' (1621) by Stefano Bernardi is not an original work but a skilful arrangement of the composer's five-part madrigal 'Bellezze amate' (1619).[13] As there are only two roles (Sponsus and Sponsa), the remaining parts are allotted to various instruments (violin, cornetto, and theorbo); their ritornellos detract to a certain extent from the spontaneity of the conversation.

Although a short piece by Giovanni Ghizzolo ('Indica mihi' (1615)) is not called a *dialogo*, it nevertheless presents exchange of speech. A noteworthy trait in the melodic style is the frequent occurrence of conflict between rhythm and metre.

The secular character of the Song of Songs posed a problem to the church authorities: how to bring this biblical book, glorifying physical love, in harmony with the spirituality of the Scriptures. The solution, already adopted during the Middle Ages, was that of a metaphorical interpretation. The love of the Sponsa is (anachronistically) directed to Jesus (the Sponsus). Once this was accepted, the unabashed eroticism was no longer felt disturbing. The texts of several Canticle dialogues underline the spiritual message.

Like 'Tolle, sponsa, tolle fores', the piece entitled *Sponsa Canticorum*, scored for three sopranos and bass with two violins, was incorrectly attributed to Carissimi. Its freely invented text, including a few scriptural quotations, transforms the scene of the girl begging assistance from the Daughters of Jerusalem into a conversation between the male lover and the group of women. Knowing that his beloved suffers, he urges them to go to 'the mountain of myrrh', to offer her lilies, roses, and honey, and to console her. Although this dialogue contains little action, it presents a lively exchange, in particular because the three sopranos also appear individually. The vocal and instrumental ritornellos following each of their solo sections contribute to the structural coherence of this attractive composition. The spiritual implication becomes clear only towards the end, when the Sponsus, echoed by the group of women, sings:

> Who can know me and not seek me?
> Who can withdraw and not miss me?

These words point unequivocally to the Sponsus being Jesus and the Sponsa representing a soul in search of her Saviour.

Another dialogue that initially conceals its spiritual message is 'Adjuro vos, filiae Ierusalem' (1612) by Giacomo Finetti, a prolific and celebrated composer of sacred music (his printed motets were republished in seven volumes in

---

[13] S. Bernardi, *Il terzo libro de madrigali a 5 voci* (Venice, 1619).

Frankfurt). As usual, the piece is scored for four voices, the soprano taking the role of the Sponsa and the alto, tenor, and bass those of the Daughters of Jerusalem. Nothing remarkable happens until bar 126; questions and answers are exchanged in accordance with the biblical text. However, after the girl's eulogy of her lover, lyricism suddenly yields to moralization. The tenor sings an aria of only eleven bars, set to the words: 'There is no splendour, no shining beauty, no virtue that is not reflected in the glorious Virgin.' This text is repeated by the other voices in a short homophonic ensemble that during the rest of the composition serves as a refrain. Conspicuously, any textual relationship with the preceding dialogued part of the work is missing. No mention is even made of spiritual love.

The Sponsa representing an anonymous soul could be replaced by a female saint. This happens in a piece which is truly unique in the entire repertory of seventeenth-century dialogues. I am referring to 'Ecce spina, unde rosa' (1634) by the Sicilian composer Giuseppe Caruso, set for soprano and tenor. The text is worthy of quotation in its entirety.

1. (T) Ecce spina, unde rosa qua coroneris. Veni, amica mea, columba mea, conor-
        aberis.
2. (S) Quae vox auditur in terra nostra? Vox dilecti mei.
3. (T) Ecce spina, unde rosa qua coroneris.
                                Echo: eris.
4. (S) Ero pulchra et formosa, dilecte mi, sub umbra alarum tuarum.
                                        Echo: arum.
5. (T) Arum floribus, felix una coronaberis.
                                Echo: aberis.
6. (S) Abero, sed per vicos et plateas quaeram semper te, quem diligit anima mea.
            Echo: mea.
7. (T) Mea semper tu fuisti. Noli quaerere, invenisti quem quaeristi, Rosalea.
                                        Echo: ea.
8. (S) Eamus ergo, dilecte mi, quia inveni quem diligi anima mea. Quesivi et inveni.
            Echo: veni
9. (T) Veni, dilecta mea,
   (S) Veni, dilecte mi,
   (T) Veni de Libano,
   (S) Veni de Libano,
   (T) Veni, sponsa mea,
   (S) Veni, sponse mi.
   (T) Veni, coronaberis.
10. (ST) Egrediamur simul in agrum et commoremur in illis, moremur in illis,
                                        Echo: moremur in illis,
11. (ST) ... psallentes, laudantes, dicentes in saecula: Vivat Dominus, Alleluia,
12. (ST) ... iubilantes, collaudantes, exclamantes et dicentes: Vivat in aeternum, Alleluia,
13. (ST) ... iubilantes, collaudantes, exclamantes et dicentes: Vivat in aeternum.
        Alleluia!

This text, with its ingenious echoes anticipating the first word of the next verse (a well-known device in vernacular poetry of the time), transforms the original love-scene into a dialogue of Santa Rosalia, patroness of Palermo, whose feast is celebrated on 6 September, and Jesus. However, we have to wait until the seventh strophe before the saint's identity is revealed.

Caruso set the text as a series of thirteen strophic variations on a pre-existing bass, that of the *pavaniglia* (related to those of the *pavana hispanica* and the *folia*). The strophes 1–8 are alternately sung by the two voices, following which they exchange short phrases (9); finally, they sing together (10–13). The expert use of Lombardic rhythm in the melodies confers on the piece a particular charm. Here we have an example of a minor composer capable of writing music that is not only original but also of high quality (see the transcription in Part Two).

Other works do not conceal at all the spiritual interpretation of the scriptural words. The Canticle dialogues by both Michel'Angelo Grancino and Domenico Mazzocchi reveal their sacred character through textual additions in the opening sections. Thus Grancino's first phrase reads: 'Quid est quod dilectus meus *angelicus* loquitur mihi?' Hence the invocation of Jesus towards the end of the piece does not come as a surprise. This dialogue is remarkable because of the addition of an optional violin part. The instrument first functions as a counter-voice to the vocal bass; in the final duet it is allowed a few idiomatically virtuosic passages.

Unlike most of his other dialogues, Mazzocchi's *Dialogo dell Cantica* was not written for services in the S. Marcello oratory, the subject being unsuitable to the Lenten period. The words are freely based on verses 5: 8ff., but the answer of the Sponsa to the first question of the Daughters of Jerusalem leaves no doubt about the identity of the absent lover: 'Amor meus candidus et rubicundus *crucifixus* est.' The text reads almost like an allegorical representation of the scene of Mary Magdalene at the tomb of Christ.

This work shows once more Mazzocchi's great ability as a composer of dialogues. Written for a solo soprano and a four-part ensemble, it contrasts the characters not only in scoring but also in style. The musical language of the Sponsa contains unusual rhythmic and harmonic features (see Exx. 22, 23), depicting her emotional state, which are not found in the passages of the ensemble.

Although to a modern mind the traditional interpretation of the Canticles as referring to the union of a human soul with Christ or the Virgin may seem a mental *tour de force*, during the seventeenth century it was readily accepted. An era which produced works of art like Bernini's sculpture of St Teresa had no difficulty in elevating eroticism to the level of spirituality. And so the Song of Solomon could rightly be called a *liber didacticus*.

Ex. 22

Ad - ju - ro vos, fi - li - ae Je - ru - sa - lem,

num quem di - li - git a - - ni - ma me - a vi - di - stis?

Ex. 23

A - mor me-us can - di - dus et ru - bi - cun - dus cru - ci - fi - xus

est, et i - pse est di - le - ctus me - us

3

Italy: Dialogues Based on the
New Testament

The Annunciation and Nativity

The Annunciation of the birth of Jesus as recorded by St Luke (1: 26–38) was a favourite subject of the composers of dialogues during the first three decades of the seventeenth century. In general, scriptural text describing the conversation between the archangel Gabriel and the Virgin was set in essentially unaltered form. Broadly, one can distinguish three different types of setting. The first is for two voices only, the Virgin singing soprano and Gabriel countertenor, tenor, or (rarely) bass. Because of the absence of a *historicus*, the portions of narrative text are either omitted or changed into direct speech. Verse 29, for example ('And when she saw him, she was troubled at his saying, and cast in her mind what manner of salutation this should be'), was reduced to: 'What is this salutation?' Among the Annunciation dialogues of this type there is a short piece, 'Ave gratia plena' (1618), by the Roman composer Girolamo Bartei. Written in an extremely sober style, it contains only the indispensable parts of the conversation. Neither the name of Jesus nor his future kingdom as related in verses 32–3 are mentioned. The Virgin's final words (v. 38: 'Ecce ancilla Domini, fiat mihi secundum verbum tuum') are sung first solo and then by both voices as a *conclusio*.

An anonymous dialogue contained in *Scelti di motetti* (1618) was probably written also by a Roman composer, since the collection was published in that city.[1] The text is almost identical with that of Bartei's piece. Yet one small detail enlivens the scene. Following the announcement that she will conceive and bring forth a son, the Virgin's words 'How shall this be, seeing that I know not a man?' are interrupted by Gabriel: 'Listen, Mary'; only then is she able to complete her question (Ex. 24). The final duet uses the words of verses 32 and 33, omitted in the dialogue proper: 'the Lord God shall give unto him the throne of his father David: And he shall reign over the house of Jacob for ever; and of his kingdom there shall be no end.' This produces a convincing conclusion for the scene. Taken as a whole, this anonymous work is particularly unpretentious. Yet

1 Anon., 'Ave gratia plena'.

Ex. 24

it is precisely this quality which shows its consummate craftsmanship. A lesser composer would have failed to hold the listener's attention by such simple means.

Giovanni Francesco Capello treated the subject in an affective style.[2] Typical of this composer is the differentiation between the two roles. The words of the Virgin are set with more forceful expression, involving chromaticism and syncopated rhythms, than those of the archangel. Yet in the latter's part the mention of the name of Jesus is highlighted by a sudden slowing down of the melodic rhythm (Ex. 25). The *conclusio* is again different. Based on a freely invented text, it states simply that the scene has come to an end: 'Ecce completa sunt omnia quae dicta sunt per Angelum de Virgine Maria.'

Even more affective is the musical language of Adriano Banchieri. His 'Ave Maria' (1625) contains several melismatic cadential passages that remind us of the Florentine style. These, however, being unrelated to the words, do not contribute to a heightened emotional intensity. In this respect the more sober dialogues discussed above produce greater effect. The opening phrase of Banchieri's piece (Ex. 26b) borrows the melody of a psalm antiphon for Second Vespers on the feast of the Annunciation (25 March) (Ex. 26a).

Such dialogues are for two parts. A second type adds a separate part for the *historicus* or *testo* to those of the 'actors'. In some instances the narrative text

[2] G. F. Capello, 'Ave Maria' (1610).

Ex. 25

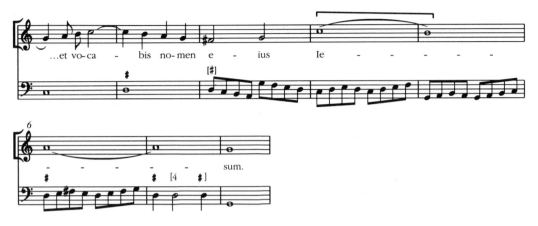

portions are taken from Vesper antiphons rather than from the Scriptures. The closing section is invariably set for five voices.

Two great masters of the time, Alessandro Grandi and Ignazio Donati, wrote their Annunciation dialogues in this manner. The text of Grandi's 'Missus est Gabriel' (1610) is exceptional in presenting the thirteen verses from St Luke's Gospel without any omission. The setting of the roles for Mary (A), Gabriel (low T), and the 'Texto' (B) contrasts with that of the two celestial sopranos in pitch as well as through its slower rhythm. These 'concealed' voices (already mentioned in Chapter 1) lend the piece an additional dimension. The sopranos sing their antiphon text either between the phrases of the soloists or simultaneously with them. They reappear in the conclusio, albeit with different words. By this happy contrivance and its most skilful application Grandi succeeds in raising the scene to a level that involves both heaven and earth.

Donati's piece (1618), which offers no celestial comment on the scene, provides a more florid setting of the scriptural words. The composer shows his great melodic gifts, especially in a few fragments that are repeated with heightened emotion. The final section, written in transparent counterpoint, is set, significantly, to the famous words 'Et verbum caro factum est' (John 1: 14).

These two masterly works, both included in Smither's anthology, belong to the most remarkable dialogues written in the seventeenth century.

The same cannot be said of Natale Bazzino's 'Angelus Gabriel descendit' (1628). Yet this short piece, which takes as its subject the announcement of the birth of John the Baptist, is not at all unattractive. The melodic style shows some Florentine influence and a few melismatic motifs clash delightfully with the harmony. The text, based on Luke 1: 11–13, 57–62, and 68, treats both the announcement and the birth succinctly. The omission of verses 18–20, containing the angel's verdict on Zachariah's remaining dumb until the birth of his son,

Ex. 26*

is rather awkward. No explanation is given of the fact that the father has to write down the name of John. These words, which, strictly speaking, belong to the role of the *historicus*, are nevertheless sung by Zachariah: 'Johannes est nomen eius'. In this case realistic presentation yields to emotional intensity, the phrase in question containing a long melisma on 'eius'. Subsequently, the father's tongue is loosed and the opening verse of his hymn, serving for the concluding section, is sung by all five voices. The counterpoint of this ensemble compares unfavourably with that of Donati's piece; it lacks the latter's flexibility and sounds rather stiff.

The third type of Annunciation dialogue differs from the second only in that it represents the narrator by a five-part ensemble. Examples include works by Severo Bonini (1609) and Giovanni Francesco Anerio (1619). Bonini's piece shows a contrast between the Florentine style of the solo parts and the rather conservative treatment of the ensembles. As for Anerio's dialogue, here the ensembles predominate to such an extent that they impair the composition's structural balance. Curiously enough, the angel speaks only about the Holy Ghost and 'the power of the Highest' that shall come upon and overshadow the Virgin; no explicit mention is made of the forthcoming birth of Jesus. The listener has to wait for the concluding five-part ensemble to be enlightened on this point.

Finally, there is a double-choir dialogue, 'Exurgens Maria' (1618) by Giovanni Battista Stefanini, the text of which relates to both the birth of John the Baptist and that of Jesus. The event is the Virgin's visit to her cousin Elisabeth. The composer set the words, freely elaborating Luke 1: 39ff., explicitly in Roman style (see Chapter 1). The two four-part ensembles are treated antiphonally as well as united in an eight-voice body. In accordance with Roman tradition the expression of the text is restrained; only the mention of Elisabeth's babe leaping in her womb (1: 44) is accompanied by lively rhythm. The solo sections, too, are

soberly treated; there is no trace of affective melodic writing. The work concludes with the opening verses of the Magnificat. The setting of the first of these,
sung by the Virgin alone, once more provides a rare example of the use of plain-
chant, quoting a transposition of the melody in the second tone (Ex. 27).

Ex. 27

After 1630 the Annunciation practically disappeared from the dialogue repertory. There may have been several reasons for this. The tendency to paraphrase
biblical words, characteristic of the middle of the century, was perhaps considered unsuited to the subject. As regards the music, the growing predilection for
sections written in aria style did not fit the scene. It is difficult indeed to imagine
the archangel or the Virgin singing an aria. Obviously, the continuous use of semi-
recitative was regarded as the only proper means of expressing the scriptural text.

Unlike the Annunciation, the Nativity, that is, the scene of the angel and the
shepherds (Luke 2: 8ff.), does not contain a true exchange of speech. The angel,
having delivered the message of the Saviour's birth, retires, and so does the
'heavenly host'. It is only then that the shepherds say to each other: 'Let us now
go even unto Bethlehem, and see this thing which is come to pass, which the
Lord hath made known unto us' (2: 15). As it proved virtually impossible to
write a true dialogue with strict observance of the scriptural text, the Nativity was
treated much more freely than other biblical subjects. Various details were
invented, and the order of the events was sometimes changed, as, for instance,
the chorus 'Gloria in excelsis Deo' preceding the angel's words. As for the music,
although solo parts are not lacking altogether, we find throughout the century a
predilection for alternating groups of voices (angels and shepherds). This
offered the composers the opportunity to employ contrapuntal devices (canonic
passages) and homophonic refrains ('Gloria' or 'Noé'). The use of semi-recitative
was restricted to small portions of text; generally, the composers preferred to set
the words in dance-like triple time or (later) in aria style. In view of the subject's
free handling it is hardly surprising that the 'cast' of a Nativity dialogue only

rarely includes a narrator. Obviously, a narrative role was thought to detract from a spontaneous and lively presentation of the scene.

The difference between Nativity dialogues and those based on other scriptural subjects is already apparent, in the early years of the seventeenth century. Two pieces, one by Ottavio Catalani, the other by Giovanni Battista Fergusio, evince the external characteristics of the Roman style: double-choir scoring with intervening solo sections. Yet there is no trace of the austerity common to this type of scoring. On the contrary, the subject is treated with an exuberant expression of joy. This is particularly true of Catalani's 'Angelus ad pastores ait' (1616). Although his piece includes a minor narrative role (accounting for only twenty-six bars out of a total of 261), the emphasis is laid on the refrain 'Noé', repeatedly sung by both the angels and shepherds. The concluding tutti section is set to the text of the angels in heaven: 'Gloria in excelsis Deo'.

In Fergusio's short dialogue (1612), set for seven voices, these words form the opening phrase. For the rest, the text paraphrases the Nativity antiphon 'Quem vidistis, pastores' (1629). In addition to profuse melismatic passages, the piece contains harmonic licences (for instance, a suspended ninth resolving on the lower fifth), features that are quite alien to the early Roman style.

Pietro Bertolini's 'O felices pastores' is another example of a dialogue freely based on the 'Quem vidistis' antiphon. Set for four parts, it presents the text in alternating pairs of voices that are treated homophonically as well as in counterpoint (see Ex. 8 above).

A dialogue by Gasparo Casati (1640) concerns the Epiphany rather than the Nativity. However, the same antiphonal opening words are used again, the 'pastores' being replaced by 'Magi'. The paraphrased scriptural text (Matt. 2: 1 ff.) is presented as conversation between an angel and the three 'wise men'. The former's questions are answered in particularly ingenious counterpoint; yet the composer took care that, by inserting homorhythmic passages, the words are perfectly understandable. Characteristically, the whole piece, with the exception of thirty-three bars, is set in triple time, and so there is hardly any trace of semi-recitative writing.

Chiara Margarita Cozzolani did not excel in contrapuntal technique. However, she was quite capable of presenting a lively scene by means of expert handling of the different roles as well as by varying the melodic style and exploiting contrasts of tempo. Thus her 'Gloria in altissimis Deo' (1650), set for four voices, gives an attractive version of the scene in the field near Bethlehem. Both the angels and the shepherds appear in pairs as well as individually before concluding the work with a quartet ('Alleluia'). (A transcription is included in Part Two of this study.)

No less attractive is a short setting found in an anonymous Como manuscript, the style of which points to the mid-century.[3] Three shepherds (ATB) are

---

[3] Anon., *Nascita di Gesù*.

opposed to two angels (SS). Instead of concluding the piece with a five-part 'Alleluia', the composer noted after the final ensemble of shepherds ('Eamus igitur') the words: 'Si suonerà la piva' (The bagpipes will play).

Bagpipes are also present in Maurizio Cazzati's 'Ad cantus, ad gaudia' (1668), albeit only in the guise of drones in the continuo part. This piece, written on a free text without scriptural quotations, features one angel and two shepherds. The variety of short contrasting sections, already apparent in the two previously discussed compositions, is pushed to extremes. In the course of the work the musical metre changes ten times. Some short sections are termed 'aria' (seven bars) or 'recitativo' (two bars). Changes of tempo, explicit or implied, occur almost as often as those of metre. Taken as a whole, this long dialogue (291 bars) is rather unbalanced; it leaves the listener almost breathless.

Cazzati's piece, dated 1668, already belongs to the final phase of the dialogue genre. *Nenia ad Iesum infantem* (1681) by Natale Monferrato was published when the composer had almost reached the age of eighty. Yet the work shows various modern traits, such as an extended strophic duet in the central section. On the other hand, the notation of the vocal *trillo* (an accelerating repeated note) reminds us of the composer's youth (he was born *c*.1603). Although, according to Brossard, this ornament was still in use during the last decades of the century, it was practically never set down in notation.[4]

Monferrato's dialogue is exceptional in dealing with both the scene in the field and the shepherds' visit to Bethlehem. The tone is particularly tender, and the central duet as well as the ensuing adagio aptly illustrate the title of the composition: a lullaby. The depiction of 'sleep' in the continuo part is ingeniously applied to both the shepherds (in the opening bars, Ex. 28a), and the child (Ex. 28b).

An invented detail enlivens the start of the scene. The second shepherd feels irritated when he is wakened by the angel ('Quae vox importuna mihi somnum obrumpit?'), whereas the first one is overwhelmed by the heavenly light ('Quae lux insolita mihi luces perstringit?'). The concluding homophonic terzet 'Noé' runs to a mere eight bars, leaving the tender character of the composition unimpaired.

Episodes from the Life of Jesus

Two early examples of dialogues based on the Finding of Jesus in the Temple by Viadana and Balbi were discussed in Chapter 1. Very soberly set, they leave hardly any room for the expression of feelings. Another setting, by Natale Bazzino, 'Video pulcherrimam mulierum' (1618), shows how little was needed

---

[4] See S. de Brossard, *Dictionaire de musique* (2nd edn., Paris, 1705; repr. 1965), article 'Trillo', pp. 169–70.

Ex. 28

to turn the rather dispassionate scriptural text (Luke 2: 43–9) into a moving scene. The composer added to the roles of Mary and Joseph one for an invented anonymous friend (Amicus). It is he who, seeing the alarmed parents in search of their son, comments on their sorrowful countenance. During the ensuing conversation both the father and mother utter feelings of distress. The setting, showing some Florentine influence, includes a striking dissonance on the word 'dolor' (an unprepared and irregularly resolved major seventh). After this the friend's information that the boy is to be found in the Temple, disputing with the scribes, comes as a true relief. Both Joseph and Mary express their gratitude in a joyful triple-time section. In the Temple the parents' mild reproof is again depicted by sharp dissonances on the word 'dolentes'. Jesus's answer takes only seven bars, and the piece, which is among the best of Bazzino's dialogues, concludes with a five-part ensemble involving the listener: 'Gaudeamus cum Maria Virgine.'

No emotion is vented in two settings of the scene of the Temptation in the Desert (Matt. 4: 1–11). Donati's 'Cum jejunasset Jesus' (1618) scrupulously follows the scriptural reading, assigning a two-part role to the Historicus. In Quagliati's 'Ductus est Jesus' (1627) the narrative words are sung by a double choir, the part of Christ and the devil being reduced to the indispensable portions of the text. Although both works offer musical interest, they do not present the scene as a dramatic confrontation of antithetical forces. The subject-matter is expressed cerebrally rather than emotionally.

The first miracle, the turning of water into wine at Cana, was treated as a dialogue by Stefano Bernardi. This piece, which is preserved in several prints and manuscripts, features Jesus (A), Mary (S), the Architriclinus (governor) of the wedding (T), and three servants (BBB). The latter's low voices form a special characteristic of the work. Hardly any other dialogue of the time has a beginning such as that shown in Ex. 29. The three basses also lend an unusual colour to the brilliant six-part concluding ensemble. Although Bernardi remained practically untouched by the modern stylistic trends of his time, one cannot deny him originality.

Ex. 29

The scene of Jesus and the Samaritan woman at Sychar (John 4: 4 ff.) provided material for dialogues by Donati[5] and Marazzoli.[6] The latter's work, an oratorio dialogue scored for seven voices with four strings and intended for the service in the S. Marcello oratory on the third Friday in Lent, shows once more the composer's restraint in rendering a biblical episode. The conversation between the woman (S) and Jesus (T) alternates with freely invented commentary sung by various ensembles. Donati's piece, lacking a narrative part in the dialogue proper, contains only the scriptural words in direct speech. Conspicuously, all the woman's questions end on weak metrical beats. The text of the 'Conclusione a 5' is based on verses 40–2, describing the conversion of the Samaritans.

Dialogues dealing with further episodes from the life of Christ were set mostly by Roman composers. They include the representation of performed miracles as well as conversations held with Pharisees, scribes, disciples, and other persons. The faithful rendering of the scriptural text, characteristic of Rome, inevitably

---

[5] I. Donati, 'Mulier, da mihi bibere' (1618).   [6] M. Marazzoli, 'Venit Jesus in civitatem Samariae'.

resulted in a rather 'dry' handling of the verbal material. Typical in this respect
are the five-part settings of various episodes by Domenico Massenzio (1616),
such as the healing of the blind beggar (John 9: 1 ff.),[7] the dispute with the
scribes and Pharisees about strict observance of the laws of Moses (Matt. 15:
1 ff.),[8] their asking for a sign (Matt. 12: 38 ff.),[9] and the mother of Zebedee's sons
claiming for the latter a place in Jesus's kingdom (Matt. 20: 20 ff.).[10] All these
pieces are of high musical value yet rather abstract in their interpretation of the
texts. A subtle detail like the ascending scale depicting the words 'Et extendens
manum' is a rare exception.[11]

The same observation can be made about most of the other composers working in the area of Rome. Their works include eight-part dialogues by Lorenzo
Ratti (1628) on the subjects of the healings of a nobleman's son (John 4:
46 ff.),[12] the palsied man let down through the roof (Matt. 9: 1 ff.),[13] and the
leper and the centurion's servant (Matt. 8: 1–13).[14] An undistinguished dialogue
about the Feeding of the Five Thousand (John 6: 5–14) was written by Vincenzo
Pace (1613).[15] The ensuing speech to the multitude (John 6: 32 ff.), set for six
voices by Agazzari (1613), includes, apart from the roles of the Historicus and
Jesus, a homophonically treated ensemble of the Turba.[16] Like so many other
works, this piece contains doctrinal statements that leave no room for the
expression of personal feelings.

Two north-Italian dialogues present discourses on the subject of renunciation.
These are 'Venite filii' (1649) by Giovanni Battista Treviso and 'Qui non renun-
tiat' (1660) by Giovanni Legrenzi. Both are set to fragments of text borrowed
from various gospels and patched together. For the rendering of a conversation
between Jesus and two unnamed disciples Treviso used a typical Baroque trio
scoring (SSB). The disciples, who sing continuously in homophony, are counterposed to Jesus. The setting offers pleasant music—perhaps too pleasant in view
of Jesus's words calling for the renunciation of all worldly ties, including the
relation with parents, brothers, sisters, and spouse. Legrenzi, too, scored his
piece for three voices (TTB), but in this instance the treatment of the disciples'
parts involves counterpoint as well as homophony. The words 'persecutionem
patimur et sustinemus' (bars 31–49) are significantly depicted by suspensions
and black notation symbolizing death. Unlike these two compositions the dialogue 'Stans autem Jesus' (1620), set by Monteverdi's brother Giulio Cesare, is
based on a coherent fragment of St Luke's Gospel (18: 40–3). The style of this
work, which includes roles for Christ (B) and the blind man (A), approaches that
of the Roman composers.

---

[7] D. Massenzio, 'Praeteriens Jesus'.   [8] D. Massenzio, 'Accesserunt ad Jesum . . . Quare discipuli'.
[9] D. Massenzio, 'Accesserunt ad Jesum . . . Magister'.   [10] D. Massenzio, 'Ascendens Jesus'.
[11] See 'Accesserunt ad Jesum . . . Magister', bars 85–9.   [12] L. Ratti, 'Erat quidam regulus'.
[13] L. Ratti, 'Ascendens Jesus in naviculam'.   [14] L. Ratti, 'Cum descendisset Jesus'.
[15] V. Pace, 'Unde ememus panes, ut manducent hi?'   [16] A. Agazzari, 'In illo tempore dixit Jesus'.

Individually identified disciples appear in dialogues by Leoni, Pietro Pace, Banchieri, and Treviso. Three of these deal with scenes of Jesus and Peter (Matt. 16: 18 and John 21: 15–17). While both Pace[17] and Leoni[18] treated the words unemotionally, Banchieri's setting (1625) shows expressive writing, Christ's opening phrase ('Petre, amas me?') being emphatically repeated on a higher pitch and provided with a melisma. In the final part of this work the rendering of the word 'aedificabo' by a descending scale of semibreves contrasts effectively with the fast rhythm of the countervoice. A dialogue by Treviso, 'Non turbetur cor vestrum' (1629), includes roles for the disciples Thomas and Philip; its subject is that of Jesus's farewell in the Paschal chamber (John 14: 1ff.).

The musical quality of the works dealing with episodes from the life of Jesus varies widely, ranging from the rather clumsy writing of Vincenzo Pace to the brilliant double-choir settings of Lorenzo Ratti. However, none of the subjects treated lent itself to the affective text interpretation found in many Annunciation and Nativity dialogues. Only a single scene offered composers the opportunity to convey upon the listener the personal feelings of the characters and the atmosphere lying behind the mere words. I am referring to one of the last miracles achieved by Jesus, related in great detail in St John's Gospel (11: 1–46): the Raising of Lazarus. It is in particular the strong emotional ties uniting the characters involved (Martha, Mary 'Magdalene',[19] Lazarus, and Jesus) that sets this scene apart from those mentioned above. As is shown by various examples, the scriptural text is explicit about this (v. 3: 'Lord, behold, he whom thou lovest is sick'; v. 5: 'Now Jesus loved Martha and her sister, and Lazarus'; v. 35: 'Jesus wept'). Another distinctive trait is the scene's culmination with Jesus's words 'Lazarus, come forth', spoken 'with a loud voice' (v. 43). No wonder that this biblical subject was a favourite choice among the composers of dialogues. Four pieces will be discussed here briefly.

'Erat quidem languens Lazarus a Bethania' was written by Marco Marazzoli for the oratory service of the fourth Friday in Lent. The work is set in his usual restrained manner. However, the ensembles of the Judei and Discipuli, partially untexted in the manuscript, enliven the scene as a whole. Sisto Reina used a paraphrased text for his dialogue 'O caelum iniquum' (1660). In addition to Martha (S), Mary 'Magdalene' (A), and Jesus (B), his 'cast' includes a part for an angel (S) that seems entirely superfluous. The sisters' complaints are expressed in emotionally charged semi-recitative over a bass in particularly slow harmonic rhythm. The comforting words of Jesus and the angel, on the other hand, are set in dance-like triple time, a procedure that is adopted by Martha and Mary in their final praise of God.

---

[17] P. Pace, 'Si diligis me, Simon Petre'. [18] L. Leoni, 'Petre, amas me?' (1608).

[19] While it is true that Mary of Bethany was certainly a different person from Mary Magdalene, it was widely believed in the seventeenth century that the two were the same.

Notwithstanding the value of these two, contrasting works, they do less justice to the subject than those by Ignazio Donati and Domenico Mazzocchi. In his 'Domine, si fuisses hic' (1618) Donati enhances the tension by both melodic and harmonic means. The repetition of melodic entities at a higher pitch is combined several times with the juxtaposition of chords a third apart, involving chromaticism with or without cross-relations (Ex. 30). The impressive declamation of the text and the beautiful melodic contours, both in the roles of the sisters and that of Jesus, mark out this dialogue as one of the composer's happiest inspirations.

Ex. 30

Presumably, Mazzocchi's setting (*Dialogo di Lazaro*) was written some twenty years later. His text, including a narrative role sung solo as well as by a combination of voices, contains many details omitted in Donati's piece. None of Mazzocchi's other dialogues can match this one in intensity and depth of expression. All the resources of music—melody, rhythm, harmony, and counterpoint—are put to the service of a particularly touching presentation of the scene. A comparison of phrases set by both Marazzoli and Mazzocchi is instructive (see Exx. 31 and 32, and 33 and 34).

The sequel to this scene, the conspiracy of the priests and Pharisees (John 11: 47–50), offered less opportunity for the expression of feelings. This subject involves groups rather than individual characters (there is only a single short phrase by the high priest Caiaphas). As a result, the setting by Mazzocchi (*Concilio de' Farisei*) provides more musical than dramatic interest. Yet the addition of a moving three-part ensemble of the 'angels of peace' and of a concluding eight-part section set to bitter words ('See how the righteous individual dies') enriches the scene. Another setting in Roman double-choir style by

Ex. 31

(Marazzoli)

Et la - cri - ma — tus est Je — sus.

[5 6] [4 ♯]

Ex. 32

(Mazzocchi)

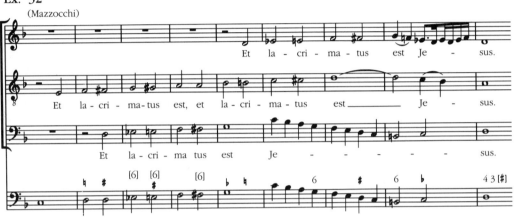

Et la - cri - ma - tus est Je — sus.

Et la - cri - ma - tus est, et la - cri - ma - tus est___ Je — sus.

Et la - cri - ma tus est Je — — — — sus.

♮ ♯ [6] [6]   [6]   ♭ ♮   6   ♯   6 ♭   4 3 [♯]

Ex. 33

(Marazzoli)

La — za - re, ve - ni fo — ras!

[6]
[5 5] [4 ♯] [♯]

Ex. 34

(Mazzocchi)

Adagio

La — za - re,   La — za - re,   ve - ni fo - ras!

♭   ♭   [6]
                          5   4 3

Alessandro Della Ciaia ('Collegerunt Pontifices et Pharisei' (1666)) testifies to the contrapuntal mastery of its composer but leaves practically no room for affective writing. The phrase of Caiaphas ('it is expedient for us that one man should die for the people and that the whole nation perish not') is repeated in the concluding section, the words 'ut unus moriatur homo pro populo' being sung by a single voice and the rest ('et non tota gens pereat') by the full ensemble.

Parables

A number of dialogues, all but one dating from the first half of the century, are based on parables. Since several of these include extensive narrative text portions, some composers set only the crucial part of the story that involves conversation. Thus Vincenzo Ugolini's 'Media nocte clamor factus est' (1619), dealing with the parable of the ten virgins, lacks the introductory descriptive verses (Matt. 25: 1–5), whose content was obviously supposed to be known to the listener. The dialogue starts with the cry at midnight: 'Behold, the bridegroom cometh.' From here on the scriptural text would lend itself well to the presentation of an animated scene, but in this instance it is the composer's conservative taste that forms an obstacle to its realization. The style of the piece is that of the late Renaissance rather than the early Baroque, the verbal exchange between the wise and the foolish virgins being written in imitative counterpoint. Only the *clausula perfecta* underlining the words 'the door was shut' and the homophonic setting of 'Lord, Lord, open to us' may be considered relatively modern traits. For the rest, this work, despite the composer's musical proficiency, misses its dramatic point.

Two interrelated parables, 'Simile factum est' (the marriage feast (Matt. 22: 2–14)) and 'Homo quidam fecit caenam magnam' (the great supper (Luke 14: 6–24)), were set by Lorenzo Ratti (1628) for double-choir including solo parts. These works contain narrative sections in which homophony and counterpoint keep each other in balance. In one instance the contrapuntal technique reflects the text faithfully. The opening word of 'Simile factum est' is depicted ingeniously by through imitation of four pairs of voices; additionally, each of these pairs features imitation in *motu contrario* applied to their initial notes (Ex. 35). Although in both dialogues the musical representation of events leading to wrath and even the slaying of men is rendered in a discreet manner characteristic of the Roman style, they contain a few examples of text interpretation. In 'Simile factum est' as well as 'Homo quidam' the rage of the indignant host is depicted by repeated short motifs in triple time, set to the word 'iratus'. Moreover, in 'Simile factum est' the men in the street, invited to the banquet, are divided not only verbally but also musically into good and bad people; this is done by the use of the major and minor third respectively.

The verbal incipit of another dialogue by Ratti, 'Simile est regnum caelorum', induced the composer to write the opening bars in a way almost identical to those of 'Simile factum est'. Yet in this case only one of the four-part 'choirs' is involved in the contrapuntal procedure. For the rest, the work shows no textual or musical affinity to the pieces previously discussed. Its subject is the parable of the debtors, whose scriptural text (Matt. 18: 23–35) is set almost completely.

Ex. 35

Ex. 35 *cont.*

Among the noteworthy details is a striking cross-relation depicting the king's rigorous command that the debtor and all his possessions be sold and payment be made (v. 25). For this and other examples of text interpretation the reader is referred to Smither's edition of the piece.

The parable of the great supper was likewise set by Tullio Cima (1625). Here the story is presented almost entirely in direct speech, as a result of which the scriptural text is drastically shortened.

Several of Jesus's parables were difficult to understand in isolation. Only knowledge of the scriptural context or exegesis by the celebrating priest could enlighten the listener. In Ratti's 'Homo quidam' the concluding 'Alleluia', which surprisingly follows the host's negative statement ('none of the men that were bidden shall taste of my supper'), points to 'the poor, and the maimed, and the lamed, and the blind' (Luke 14: 13), rather than the buyers of land or cattle (vv. 18–19), who shall 'eat the bread in the kingdom of God' (v. 15). Another parable needing exegesis is that of the householder, his vineyard, and the wicked husbandmen ('Homo erat pater familias' (Matt. 21: 33 ff.)), set by Marco Marazzoli for an oratorical service on the second Friday in Lent. This dialogue, the only one in which the composer dispensed with concertizing instruments, is remarkable for some of its solo parts and ensembles, the text of which dramatizes and elucidates the descriptive biblical words. Curiously enough, the husbandmen, unlike the servants, are represented only by a single voice.

More appealing was the parable of the prodigal son. Its emotionally charged subject-matter tempted several composers. Even the austere Massenzio was induced to express the father's joy at the return of his lost son by a long melisma.[20] While this piece ('Quanti mercenari', dating from 1612) omits the verses describing the request for the heritage portion and its dissipation, Mazzacchi's version, set for eight voices with an additional soprano part, contains practically the whole scriptural text (Luke 15: 11–32). Although less dramatic than his dialogue on the Raising of Lazarus, this work is 'spiced' with striking melodic details, such as the minor third appearing in the phrase set to 'Father, give me the portion of goods that *falleth* to me.' Its semiotic function is of course prospective (Ex. 36).

Ex. 36

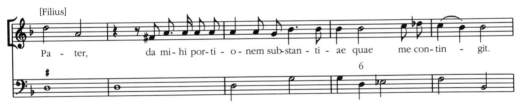

Pa - ter, da mi- hi por- ti - o - nem sub-stan - ti - ae quae me con-tin - git.

Unlike the parable dialogues so far discussed, 'O me miserum' by Giovanni Antonio Grossi dates from the second half of the century. The text of this work focuses on a turning-point in the story of the prodigal son: the hesitation and final decision to return home and ask for forgiveness. It is true that in the Bible there is no mention of hesitation; yet the implication is psychologically convincing. Two invented roles are added to that of the son (T), the personified Speranza (S) and Desperazione (A). Thus the crucial moment is highly dramatized. In the initial section the young man utters his despair, recognizing that it is only right that he should starve to death. This view is opposed by Speranza, who assures him that he may expect pity instead of punishment. Desperazione, on

[20] D. Massenzio, 'Quanti mercenarii'.

the other hand, tells the sinner that he is condemned to wail over his fate forever, since there is no hope. This section, written entirely in triple time, is followed by a similar clash of advice set in semi-recitative with melismatic endings,
especially in the parts of the two allegorical figures. Subsequently, these figures
are sharply opposed to each other in the exchange of short motifs, again set in
ternary metre (Speranza: 'Patrem adi' (Go to your father); Desperazione: 'Patrem
odi' (Hate your father)). The son finally decides to follow the advice of Speranza,
and the piece concludes with a terzet in which Desperazione persists with her
negative role. This is typical of Grossi's dialogues featuring a soul or a human
being placed between two antithetical forces (cf. the *Historia di Giobbe*, discussed in Chapters 1 and 2). Another remarkable feature of this work is the addition of trills, not only in the vocal parts but also in the continuo. These either are
purely ornamental or function as an enhancement of the emotion expressed—
as, for instance, in the son's initial exclamation 'O', which extends over more
than four bars. The piece once again demonstrates Grossi's proficiency as a composer of sacred dialogues.

The Crucifixion and Resurrection

There seem to be no dialogues dealing explicitly with events from the Passion.
However, several pieces offer conversational commentary on the Crucifixion. A
few of these are written in a form that could be called the 'interview dialogue':
an anonymous person asks the Virgin about her feelings during the sufferings of
Christ. The earliest example of this type is Giovanni Francesco Capello's 'Dic
mihi, sacratissima Virgo' (1610), already amply discussed by Howard Smither in
his 1967 article and included in his anthology. In this work the composer's application of the most modern stylistic procedures is pushed to extremes; it includes
chromatic melodic lines clashing with the harmony, unprepared dissonant intervals, and frequent instances of word-painting realized with unusual means.
However, the Virgin's final words ('En tecum morior' (See, I am dying with you))
are sung by both voices in a way sharply contrasting with the foregoing. Based
on two chords only, the canonically treated melody contains exclusively chordal
notes. Death is symbolized by the general pause in the last bar.

A comparison of this highly affective composition with a work written on a
similar text by Paolo Quagliati (1627) shows once more the essential difference
between the styles of Rome and northern Italy.[21] Quagliati set the 'interview' in a
restrained and dignified manner, characteristic of a nobleman who, apart from
being an organist and a composer, served the pope as a chamberlain. The ques-

---

[21] P. Quagliati, 'Dulcissima Virgo Maria'.

tions are assigned to a double choir that is built up gradually, the original one or two voices being added to successively until the whole eight-part ensemble is involved. The Virgin's short answers are sung by the soprano of the first 'choir', which thus changes its identity. Although, in accordance with the Roman tradition, the harmony is simple and elementary, there are two passages whose emotional text caused the composer to deviate from his restrained style: a chromatic rendering of the words 'et sic dolore consumeris' (Ex. 37a) and a striking cross-relation involving an augmented triad for the word 'occisus' (killed) (Ex. 37b). The Virgin's part includes an allusion to the Lamentations of Jeremiah (1: 12); however, its reserved treatment is quite unlike the expressive style of the same role in Capello's piece. Quagliati, too, concludes his composition with a general pause symbolizing death; moreover, the last barline is significantly omitted, since it would contradict the concept of eternity.

Some forty years later the Sienese Alessandro Della Ciaia (also a nobleman) wrote a truly stupendous piece on the same subject. This is the *Lamentatio Virginis in depositione Filii de Cruce* (1666). Its double-choir scoring reminds us of Rome; its expressive style of northern Italy. Unlike Quagliati, Della Ciaia gave a separate soprano part to the Virgin. The other voices (angels), quoting more than once the text of the 'Stabat Mater' sequence, function as commentators, rather than interviewers. The scene is opened by a narrator, whose part is sung by the two middle voices of the first 'choir'. The piece is monumental in that it displays impressive homophonic treatment of the two four-part ensembles. As for the Virgin, her text, expressing not only sentiments of utter distress but also indignation and reproach, is set in a highly original way, as shown by Exx. 38 and 39. Other striking passages can be found in the complete transcription of the work, included in Part Two. It seems almost incredible that a totally unknown composer was able to write music of such surprisingly high quality.

The text of Giovanni Legrenzi's *Dialogo delle due Marie* (1655) offers only slight dramatic interest, as it contains merely utterances of sorrow about the loss of Christ without any interaction of the two characters. However, its purely musical qualities mark out this piece as one of the most attractive in the whole dialogue repertory. Though notated with a signature of only one flat, the key is that of F minor. Within this key and its neighbours (B flat minor, C minor) the composer applies bold harmonic procedures, showing himself to be ahead of his time. A duet section, 'O Jesu dulcissime', functions as a unifying musical refrain. The overall style is very akin to Venetian operatic bel canto; it gives predominance to musical invention at the expense of early Baroque rhetoric. This composition, too, is included in Part Two.

Dialogues about the Resurrection deal mostly with the scene of Mary Magdalene at the tomb of Jesus. In view of its dramatic and emotional implications, it is hardly surprising that, in the course of the seventeenth century, this story became a favourite subject. The main scriptural source was the twentieth

Ex. 37

chapter of St John's Gospel, which relates the events in a very detailed way. Here is a summary:

1   Before sunrise Mary Magdalene comes to the tomb and sees that the stone has been removed.
2   She hurries to two disciples, one of them Peter, and tells that the Lord has been taken away.
3–10   Both disciples speed to the sepulchre; they enter and see only the linen garments. They return home.

Ex. 38

Ex. 39

11–12   Mary Magdalene stands before the tomb, weeping; she stoops, looks inside, and sees two angels.

13   The angels ask her why she is weeping. She answers that they have taken away her Lord.

14   She turns back and sees Jesus standing, but does not recognize him.

15   Jesus asks: 'Why are you weeping, woman?' Supposing him to be the gardener, Mary Magdalene asks in turn: 'If you have taken him away, tell me where you have laid him.'

16   Jesus calls her by name: 'Mary'. Recognizing him she exclaims: 'Master!'

17   Jesus answers: 'Do not touch me, for I am not yet ascended to my Father.'

18   Mary Magdalene goes to the disciples and tells them that she has seen the Lord and that he has spoken to her.

In addition, a few variant verses, borrowed from the gospels of St Luke and St Matthew, served as material for the dialogued texts. These are the angels' words spoken to Mary Magdalene and the other women: 'Why seek ye the living among the dead? He is not here, but is risen' (Luke 24: 5–6); further, the message given to the women by a single angel: 'Go quickly, and tell his disciples that he is risen from the dead; and, behold, he goeth before you into Galilee; there shall ye see him' (Matt. 28: 7).

Among the eight works to be discussed here only a dialogue by Ghizzolo covers practically the whole story as regards Mary Magdalene. The texts of four works by Fergusio, Mazzocchi, Grancino, and Cozzolani limit themselves to her conversation with the angel(s), while two pieces, both by Banchieri, focus on her confrontation with Jesus. A dialogue by Fattorini differs from all the others by dealing with the words exchanged between Mary and the group of disciples (not contained in the Bible, but freely derived from Matt. 28: 7–8 and Mark 16: 7).

The piece by Fattorini, 'Dic nobis, Maria', the text of which is borrowed from the Eastern sequence, is the earliest known role dialogue (1600). Scored for two voices only, it has a rather awkward layout, as one of the disciples shares the soprano part with Mary Magdalene. Further, the words 'Praecedet vos in Galilea' are sung by Peter instead of Mary; in view of the scriptural source, this makes no sense. Although the work is historically interesting, its musical value is low, the melodic style lacking emotional expression. Only the two-part setting of the concluding 'Alleluia' conveys joyful sentiments.

As has been said before, Giovanni Ghizzolo, in his 'Mulier, quid ploras?' (1615), treats both the scene of Mary Magdalene with the angels (SA) and the subsequent one with Jesus (B). Since the work is scored for four different types of voice, only the tenor is left for the female protagonist. This may seem odd to us, but was less so to the seventeenth-century listener, who was used to the performance of female roles by male voices (including falsettists and castrati). The text offers a condensed version of the story with a few changes in the order of the events. One non-biblical detail is Mary's complaint ('Heu me misera') after Jesus's words 'Touch me not'; her joy over the Resurrection is expressed only in the concluding quartet. Ghizzolo's piece shows that even a conservative composer was able to express the scene's emotional character; this is apparent from his setting of Mary Magdalene's part, as at the moment when she recognizes Jesus (Ex. 40).

Ex. 40

As in his other dialogues, Giovanni Battista Fergusio set the scene at the sepulchre in Roman style.[22] To the double choir representing the angels, a *canto separato* is added for the role of Mary. In another sense, too, her part is distinguished from that of the ensemble. The preface to the composer's collection of dialogues and motets states that the solo voice's *fondamento* should be

---

[22] G. B. Fergusio, 'Heu me misera'.

played on a plucked instrument (theorbo, chitarrone), whereas the other vocal parts are to be supported by the organ.

The opening is in the usual reserved style; however, in the course of the piece both the musical language of Mary Magdalene and that of the ensemble become gradually more charged with emotion. The reason for this is that, despite the angels' announcement of the Resurrection and the subsequent alleluias, the protagonist, overcome as she is by grief, twice repeats her complaint. It is only towards the end that she becomes aware of the happy event. By this contrivance the composer strongly reinforces the dramatic impact.

Another means of enlivening and enriching the scene was the importation of verses borrowed from the Song of Songs (5: 8ff.). Mary Magdalene takes the role of the Sponsa and the angels that of the Daughters of Jerusalem. We find this textual procedure in the dialogues by Grancino (1631)[23] and Cozzolani (1650),[24] written for six and four voices respectively. Grancino provided roles for two angels and Mary, the six-part ensemble being used only for the *conclusio*. The insertion of Canticle verses seems artificial, yet it enabled the composer to stress the dramatic moment in which the angels, having played a role within a role, turn to reality and call the weeping woman by name: 'Maria! In gaudium vertatur tristitia tua, quia dilectus tuus quem queris non est hic, sed resurrexit . . .' (Mary! Thy sorrow turns into joy, because thy beloved whom thou seek is not here, but is risen . . .)

Sister Cozzolani wrote a particularly attractive piece. It features a narrator (T) in addition to the protagonist (A) and the two angels (SS). The composer stresses the feelings of joy rather than those of sorrow and takes full advantage of the lyrical qualities of the Canticle verses. Several solo sections and duets are set in an aria-like style, and the *conclusio* sung by all four voices is most gratifying.

Little need be said about Domenico Mazzocchi's *Dialogo della Maddalena*. The composer's superb writing for solo voices, full of melodic, rhythmic, and harmonic subtleties, and his most expert handling of the double-choir technique have been mentioned several times. These qualities can be admired once more in the present work, which undoubtedly forms the peak of all the dialogues dealing with the scene at the sepulchre.

It is a curious fact that, apart from Ghizzolo, only one Italian composer treated the highly dramatic confrontation of Mary Magdalene and the risen Jesus. In 1607 Adriano Banchieri set this part of the scene a 4, Mary being represented by three voices (CCT) and Jesus by a second tenor.[25] Eighteen years later he returned to the same subject-matter, this time limiting the scoring to two voices (SB).[26] A comparison of the two versions is most instructive.

[23] M. Grancino, 'Quid dicam?'   [24] C. M. Cozzolani, *Dialogo fra la Maddalena e gli angeli*.
[25] A. Banchieri, 'Mulier, cur ploras hic' (1607).   [26] A. Banchieri, 'Mulier, cur ploras hic' (1625).

Although in the first piece the homophonic setting serves the intelligibility of Mary Magdalene's words, the representation of a single character by three voices restricts the freedom of expression. Yet, at the crucial moment when she recognizes Christ, Banchieri succeeds in conveying her emotion by the repeated exclamation 'O' (Ex. 41). In the later version, these outcries recur but now simultaneously with the words of Jesus: 'Dost thou not recognize me, Mary?' This procedure proves much more effective (Ex. 42). In other respects, too, the dialogues of 1607 and 1625 show strong differences. While it is true that Banchieri was familiar with Florentine monody as early as the first decade of the century, the three-part setting of Mary's role precluded the application of its stylistic properties. The scoring of the 1625 version, on the other hand, did offer the opportunity to depict her sorrow and subsequent joy by means of the *stile moderno*. This also implied the use of the *seconda prattica* (Ex. 43).

Ex. 41

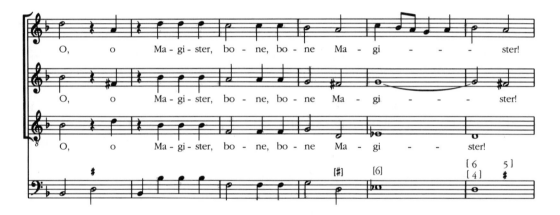

Ex. 42

Ex. 43

...et ne - - sci - o (ubi posuerunt eum)

The predilection for the setting of the scene at the tomb of Christ in dialogue form was not restricted to Italy. In Chapter 4 we will discuss other examples from England, the Netherlands, and France.

Three composers treated the episode of two disciples meeting Jesus on the way to Emmaus (Luke 24: 13–31). These were Giovanni Battista Vacchelli, Natale Bazzino, and Giacomo Carissimi. The Minorite Vacchelli, who worked as organist in Bologna and *maestro di capella* in Pesaro, used a condensed text ('Qui sunt hi sermones?' (1667)). As a result the conversation is deprived of its coherence; Jesus's exclamation 'O fools, and slow of heart to believe all that the prophets have spoken' (v. 25) refers to the disciples' previous words 'But we trusted that it had been he which should have redeemed Israel' (v. 21); yet in the text of the composition this sentence is lacking. Musically speaking, the dialogue proper is of little interest, but the concluding 'Alleluia, surrexit Dominus', sung by all three voices and accounting for more than half of the work, contains attractive contrapuntal writing. Bazzino's 'Duo discipuli ibant in castellum' (1628) suffers from the same shortcomings caused by the use of an incomplete text. However, towards the end the scriptural words are ingeniously reinterpreted. The disciples' invitation addressed to Jesus 'Abide with us, for it is toward evening (and the day is far spent)' is set as a five-part *conclusio*, adopting in this way a meaning that transcends the story proper: a prayer for divine assistance during the evening of life. The addition of a final bar containing the symbol of death (a general pause) is quite significant in this respect.

Carissimi's 'Duo ex discipulis' is set to words that closely follow the indispensable portions of the scriptural text. There are parts for the two disciples (SS) and Jesus (T); the latter's voice is used also for the narrator. Three times in the course of the piece the singers abandon their roles and together perform a terzet set to non-biblical words (twice 'Ite felices, ite beati' and the concluding 'Eamus, surgamus'). The first, repeated, terzet is written over a freely treated basso ostinato. Its triple-time bel canto style, showing Venetian influence, is adopted by the disciples. Because of this Jesus's words 'O fools . . .' need not be emphasized; they are soberly set as a recitative which strongly contrasts with the foregoing *cantabile* duet of the sopranos. The same effect is achieved at the moment of recognition and the vanishing of Jesus, related by the Testo. The final terzet shows florid vocal-writing and forms a brilliant conclusion to a beautiful work whose overall character is lyrical rather than dramatic.

There is finally a five-part dialogue about the convincing of the doubting disciple Thomas.[27] Its composer, Abundio Antonelli, was probably associated early in the century with the Crocifisso confraternity of S. Marcello. Yet this piece cannot possibly have been written for an oratorical service, its subject being unsuited to the Lenten period; the same is true of other Roman Resurrection dialogues discussed above (Mazzocchi, Carissimi). Antonelli adhered scrupulously to the biblical text (John 20: 27–9), setting it in the stately manner characteristic of his city. The narrative words are assigned to the full five-voice ensemble. In the part of Christ there is a notated cadential ornamentation in semiquavers; other florid passages may have been improvised, not only in this part but also in that of Thomas.

The Acts of the Apostles and the Revelation

Only a few dialogues deal with the Acts of the Apostles. 'O quae monstra' by Carlo Donato Cossoni (1665) is a lively scene connected with the descent of the Holy Spirit on the day of Pentecost. The scriptural text (2: 1 ff.) tells us how the disciples, being suddenly able to speak and understand the tongues of foreign people, become the object not only of amazement but also of mockery on the part of the citizens of Jerusalem: 'These men are full of new wine' (v. 13). Then follows a long discourse by Peter, as the result of which three thousand souls are baptized (vv. 14–41). In Cossoni's piece the subject-matter is transformed into an argument between an anonymous citizen (T), who accuses the followers of Jesus of drunkenness, and the apostle (B), who explains the divine effect of the Holy Spirit. After the opening antithetical declarations of the two characters ('O quae monstra'—'O stulti') a lively exchange of speech ensues:

HOMO.	Quis vos inebriavit?	Who made you drunk?
PETRUS.	Ubertas domus Dei.	The fecundity of the House of God.
HOMO.	Vos estis ebrii!	You are drunk!
PETRUS.	Sed Sancto Spiritu.	But from the Holy Spirit.
HOMO.	Vino calentes . . .	Glowing from wine . . .
PETRUS.	Deo serventes . . .	Adhering to God . . .

In a short Adagio-section the man yields to the divine power: 'Thou hast won, O Holy Spirit, and hast conquered my rigid heart.' The words 'durum cor' are significantly underlined by an augmented triad (Ex. 44). The work ends with an ecstatic song of praise set in gradually accelerating tempo (*Allegro—Presto—Prestissimo*).

---

[27] A. Antonelli, 'Jesus dixit Thomae' (1616).

Ex. 44

The rescue of Peter (12: 7–8, 11) is the subject of an eight-part dialogue by Giovanni Battista Fergusio (1612).[28] Having been thrown into prison, the sleeping apostle is smitten on the side by an unknown person. His chains fall off and the iron prison gate opens of its own accord. Believing at first that this is all a vision, Peter then realizes that it was the angel of God who delivered him out of the hand of Herod. Fergusio's piece contains roles for the narrator (T, first choir), the angel (S, first choir), and Peter (T, second choir). The composer handles double-choir technique in a most expert way. The falling of the chains, in particular, is depicted suggestively by high voices to which the lower ones are gradually added. The texts of both Peter and the ensembles include a few psalm quotations. This work was very probably written for the Second Vespers on 1 August (*festum S. Petri ad vincula*).

Other dialogues derived from this biblical book deal with the conversion of Saul (9: 1 ff.). Antonio Francesco Tomasi set this scene in a very sober way. His 'Saule, quid me persequeris?' (1615) restricts itself to the spoken words between Jesus and the future apostle, followed by a two-part 'Alleluia'. Despite its brevity this work is impressive in that it centres merely on the dramatic event related in the Scriptures. The florid passages contained in the melody set to Jesus's text 'Arise, and go into the city' are stimulating rather than affective; they serve the awakening of Saul from his stupor.

A dialogue a 3 by the Syracusan musician Andrea Rinaldi (1634) opens with the same words. This work is written in a rather old-fashioned and mannerist style, its melodies abounding in syncopated notes that give weight to the 'strong' syllables through length rather than metrical position. Characteristic in this respect is the setting of the phrase 'It is hard for thee to kick against the pricks' (Ex. 45). In addition to the scene on the way to Damascus, the work contains the summons of Anasias (vv. 10 ff.). The concluding words of Christ, 'for he is a chosen vessel to me, to bear my name before the Gentiles and kings and the children of Israel', are sung by all three voices; this is another old-fashioned characteristic of the piece.

Like most of his biblical dialogues (Latin as well as Italian), Anerio's 'Saulus adhuc spirans' (1619) shows a quantitative preponderance of the narrative and

---

[28] G. B. Fergusio, 'Angelus Domini'.

Ex. 45

reflective portions of text, which are set for the full five-part ensemble, over the action proper, which is set for soloists. In the present piece the ratio is about 3:2. The musical language remains neutral throughout the composition—no attempt is made to enhance the dramatic implications. The concluding homophonic ensemble 'Magnus sanctus Paulus' borrows its text from the Alleluia versicle of the feast of St Paul's Conversion (25 January).

The most interesting treatment of the subject is that by Fergusio (1612). We have seen on previous occasions how this composer enlivened the Roman style through the application of various procedures (see his Nativity and Resurrection dialogues, discussed earlier in this chapter). This also occurs in the dialogue dealing with the turning-point in the life of St Paul, 'Venite, congregamini'. Although this work is shorter than Anerio's setting, its textual substance contains much more. The piece even cuts across the division of the repertory adopted in this book, since it combines the past with the present, that is, the biblical story with the contemporary celebration of the saint's conversion. Thus the double choir successively adopts three different roles: it first represents the persecutors of the followers of Christ—among the latter is St Stephen—then a host of angels commenting on the conversion with quotations from Psalm 32 and the Song of Moses (Deut. 32: 43), and finally an anonymous group of worshippers celebrating the saint with a text that includes such words as 'in insigni die huius solemnitatis' (on the illustrious day of this solemnity). While in Anerio's piece the preponderance of the narrating ensembles somewhat weakens the dramatic impact of the event proper, Fergusio achieves the opposite. In particular, the opening section, in which Saul incites his fellow men to persecute the Christians (cf. 8: 3 and 9: 1), forms a striking contrast with the voice of Jesus coming from heaven.

The Revelation describes scenes of the most diverse kinds: pictures of splendid ceremony alternate with those of ghastly disaster. Yet there is hardly any verbal interaction between individual persons or groups in this biblical book. Hence the musical representation of an episode often takes the character of a sonorous tableau including, next to narrative sections, solemn statements, such as the praise of God by a group of angels. Although the description 'non-conversational dialogue' given to this type of composition in the first chapter of this book is actually a contradiction in terms, its use is historically justified, since the relevant works were all printed with the subtitle 'dialogo'.

A piece by Giovanni Battista Rocchigiani (1632),[29] representing the celestial scene described in 7: 9–12, features the narrator (John the Divine), a large crowd 'standing before the throne and before the Lamb', and a group of angels 'round about the throne, falling on their faces and worshipping God'. The scoring for only three voices precludes a clear musical differentiation of the roles. The narrative text is sung first by one and then by two sopranos; the words of the crowd as well as those of the angels are assigned to a trio of two sopranos and a bass. The composer has tried to animate the scene's static character by inserting florid passages in the part of the Testo. These do not depict anything in particular, except in the description of the angels' position: the words 'circuitu throni' are rendered by the musical figure of a circle (Ex. 46; cf. the sheaves in Joseph's dream described in Chapter 2).

Ex. 46

A similar scene, based on Rev. 15: 2–4 and 19: 6 with allusions to 1 Chr. 16: 31 and Isa. 49: 13, was set by Domenico Mazzocchi (*Dialogo dell'Apocalisse*). Here the angels (SSS), 'having the harps of God', sing the Song of Moses and that of the Lamb. The scoring of the heavenly host is for double choir: one ensemble of five, the other of four, voices. Mazzocchi's brilliant compositional technique suits the subject perfectly; it enables him to convey the ceremonial splendour by mere sound. The same text was set for five voices by Angelo Berardi (1669).[30] As in Mazzocchi's piece, the words of the angels are sung by a trio (SST) and those of the crowd by the full ensemble. As might be expected from a man who was a prominent musical theorist in his time, the work contains technically skilful writing; apart from that, it offers little of interest.

[29] G. B. Rocchigiani, 'Vidi turbam magnam'.   [30] A. Berardi, 'Stabant angeli'.

These three compositions present celestial scenes in almost strict conformity with the biblical source. One apocalyptic subject was treated differently, however: the Fall of Lucifer. The dialogued representation of this event became very popular from the middle of the century onwards. Its biblical text (12: 7–9) reads as follows:

7. And there was war in heaven: Michael and his angels fought against the dragon; and the dragon fought and his angels.

8. And prevailed not; neither was their place found any more in heaven.

9. And the great dragon was cast out, that old serpent, called the Devil, and Satan, which deceiveth the whole world: he was cast out into the earth and his angels were cast out with him.

On the basis of this succinct account, several fictitious dialogues were written; these involved not only St Michael and Lucifer, but in some cases also the angels on either side. Although the subject called for ample vocal scoring, the restricted number of voices available in Italian churches compelled a few composers to represent the scene with only two characters. One example is Orazio Tarditi's 'Surgite, o angeli mei' (1670), a conversation between Michael (S) and one of his angels (B). The former opens the piece with a call to arms, explaining the reason with a vivid description of Lucifer's blasphemous aspirations (the setting of the words 'super astra caeli exaltare solium suum audet' includes a melisma ascending to g"). This is answered by the angel representing the celestial host: 'Eamus . . . exterminemus Luciferem'. Michael repeats his incitement, singing in the unusual musical metre of 12/16 (the continuo part, though provided with the same time signature, is actually written in 6/8). The battle itself is left to the imagination of the listener; what follows is the celebration of the victory. Among the means employed by the composer to enliven the scene are long melismas, martial motifs, and contrasts of tempo.

Another two-part work concerns not the battle but its cause and outcome. This is an 'interview dialogue' by Giovanni Antonio Grossi,[31] in which an anonymous person asks Lucifer to tell him the reason for his fall: 'Once you were whiter than snow, now you are black as charcoal.' After repeated promptings Satan explains the matter and even gives his interviewer a piece of kind advice: 'Forswear pride or else you'll suffer with me in the depths.' Since Lucifer speaks about his original intention to ascend the throne of God as well as his being thrown into hell, the highest and lowest notes of his vocal range are used; in the present case the compass is conspicuously large, stretching from _C_ to _d'_. Written almost entirely in triple time, the piece is very attractive, not least because of the involvement of the continuo in the motivic interplay.

Dialogues on this subject scored for more than two parts include examples by Della Ciaia, Ghezzi, and again Grossi. Alessandro Della Ciaia shows his astonish-

---

[31] G. A. Grossi, 'Quomodo cecidisti'.

ing versatility by setting *Michaelis victoria et Luciferi casus* (a 4; 1666) in a manner quite different from that of the Virgin's lament discussed earlier. Neither his highly affective melodic style nor his unusual chord progressions are found in this piece (although in bar 71 there are two consecutive seventh chords depicting the devil's blasphemy). Instead, the tone is martial and the contrapuntal writing particularly brilliant. The opening words paraphrasing the scriptural text (v. 7) are sung by the middle voices, representing the Historicus, who, far from relating the mere facts, shows himself deeply involved in the action. This is particularly evident at the moment when he joins the archangel in the address to God: 'Tu solus Dominus, tu solus Deus, tu solus altissimus' and in the ensuing violent expressions of vengeance: 'Judica, dissipa, vindica blasphemantes te.' In the first half of the composition there are extended solo sections for each of the two antagonists in which they state their respective positions and incite their fellow angels to fight. The outcome of the battle is commented upon in a quartet ('Judicavit Dominus'). From this point the text becomes rather unsatisfying. Michael's role changes without apparent reason into that of an interviewer, his words being identical with those from Grossi's piece: 'Quomodo cecidisti de caelo, Lucifer?' (How was it that you fell from heaven, Lucifer?) Still more disturbing is the repeat of the terzet 'Judica, dissipa, vindica'; after the devil's defeat this makes no sense. However, the listener receives ample compensation for the textual deficiencies; the four-part contrapuntal *conclusio*, in particular, contains some superb music.

The most spectacular enactment of the scene is undoubtedly that of Giovanni Antonio Grossi (*Battaglia di S. Michele Archangelo col Demonio in Cielo*). The scoring for eight voices, divided into two antagonistic 'choirs', offers the opportunity to convey the impression of a true battle, as far as this is possible in a work for voices alone. The roles of the archangel and Lucifer are sung by the outer parts, the first soprano and the second bass respectively. After the devil's opening challenge and Michael's retort there follows a heated exchange of speech, the distance between verbal action and reaction becoming gradually smaller. Then the loyal angels take the offensive, expressing themselves with musical imagery: 'Ad arma! Resonent tubae, demus tympana!' The rebellious group answers with the same pugnacious words and from this point the battle is depicted with simple yet effective means (Ex. 47). In this turmoil of war there is no longer any place for individual statements. Having occupied the scene, the two four-part ensembles maintain their identity until the very end, one celebrating victory and eternal joy, the other lamenting defeat and eternal suffering.

Naturally, the means employed for the 'scenic' effect of this piece are not particularly subtle. Another martial dialogue by Grossi, likewise involving the devil and his fellow demons, offers more purely musical interest. This is 'Resarate claustra inferi', set for four voices. Its text, dealing with the Resurrection of the Dead, is vaguely based on Rev. 20. The work opens with the command of Jesus

Ex. 47

(T) to unbolt the gates of hell and to break the chains of Death. Satan (B) then summons the crowd of demons and incites them to fight, so that Christ will fall a second time (these words are depicted by descending motifs outlining triads). However, the following words sung by Jesus:

> Cede, Lucifer, recede!
> Desperata vera sors,
> profligata iacet mors
> sub potenti meo pede.
> Cede, Lucifer, recede!

prove decisive: the demons are forced to accept their defeat. The closing section opens with the antithetical phrases 'To the stars, ye souls, to life!' and 'To hell, ye demons, to death!', set to different melodies (Ex. 48). However, despite the textual divergence, the same melodic material is used further on for all four voices. Obviously, the application of compositional procedures prevailed here over individual characterization.

Ex. 48

The last dialogue about the Fall of Lucifer, Ippolito Ghezzi's 'Bella, pugna, tu cor meum', was published as late as 1708. In addition to the voices of Michael (S) and Satan (B), this work includes two independent violin parts that prove particularly important for the rendering of the martial atmosphere. This is already apparent in the introduction to the opening duet (Ex. 49). The duet

Ex. 49

proper exemplifies the idiomatic changes that took place in the years around
1700. Whereas, during the greater part of the seventeenth century, instruments
employed in motets and dialogues adopted the vocal idiom, now the tables are
turned, the voices imitating the 'language' of the violins. The result of this historical development, undoubtedly influenced by the Bolognese concerto-style, is
clearly apparent in the present work, for instance, in the passage shown in Ex.
50. This example is taken from one of the two arias included in the work; these
are written in da capo form and accompanied by the violins in unison. The
opening duet also follows this formal scheme. Only the final duet is through-
composed; Satan's admission of his defeat ('Vicisti signifer!') and Michael's cry of
victory ('Cessisti, Lucifer!') form the conclusion of the dialogue proper.

Ex. 50

Subsequently, both voices, abandoning their roles, pronounce a simple moral statement: 'This is the punishment of those who sin: they perish in the abyss.'

Although Ghezzi's piece still contains elements of the seventeenth-century dialogue, such as the inclusion of short sections written in semi-recitative, its style and formal design approach those of the dramatic cantata. So it is hardly surprising that this composer was the last to contribute to a repertory spanning over one hundred years.

4

Italy: Non-Scriptural Dialogues

Saints

With the exception of Canticle and psalm dialogues, most of the works discussed in the two previous chapters were intended for performance on specific days. The same is true of hagiographic dialogues. Generally, the feast celebrating the saint in question can easily be found in the ecclesiastical calendar. Only in a few instances does the destination remain unclear. The texts of a number of pieces written in honour of the Holy Virgin do not point to a particular Marian feast. This was probably done on purpose; it allowed performance on more than one occasion. Another, more radical, means to open-endedness was the use of a text celebrating an unspecified or anonymous saint.

Besides the scriptural saints, the hagiographic repertory ranges from early martyrs, like St Lawrence, to more recently canonized persons, such as the champion of the Counter-Reformation, St Charles Borromeo. Dialogues dealing with saints from the Bible contain expressions of praise and adoration rather than accounts of scriptural episodes. It is for this reason that they are discussed in the present chapter.

All the works are set to freely invented texts with occasional biblical quotations. Despite a few borderline cases they can be subdivided into three groups: (*a*) dialogues of praise; (*b*) those dealing with the redemption of human beings through the saint's mediation; (*c*) dramatized stories from the lives of saints, in particular their martyrdom. Strictly speaking, only the last sub-category can be called truly hagiographic.

Songs of praise are, of course, totally undramatic. Their dialogued presentation is almost always artificial: a human being asks various questions (or repeats his initial question several times in the course of the piece), enabling in this way an angel to extol the saint's outstanding virtues and describe his or her eternal triumph and joy in heaven. The lack of tension in the texts did not prevent composers from writing attractive music. This is particularly true of Chiara Margarita Cozzolani. In her setting of 'O caeli cives' (1650), a work in honour of St Catherine of Siena (30 April; see the complete text on p. 15), she employed several musical procedures to enliven the words, opposing heaven to earth by means of three sopranos (Angeli) and two tenors (Homini). Madrigalisms, such as semiquaver melismas on the word 'volate', and short melodic imitations within the separate groups of voices, not only counterpoise the monotonous text but also make the concluding homophonic section the more impressive.

Sister Cozzolani's gift of transforming a conventional text into a fascinating 'scene' is further demonstrated in 'Psallite superi', a Marian dialogue written for the feast of the Assumption (15 August) and published in her 1642 collection. This work is for two sopranos and two altos, a scoring that suggests performance in a convent. It opens with a delightful dance-like terzet, which, by being repeated three times in the course of the composition, assumes the function of a ritornello. The first alto acts as a questioner and joins the other voices only in the final restatement of the refrain. Questions and answers invariably start with the words 'Quae est ista . . .', and 'Maria est . . .', respectively. This enables the composer to use fixed opening motifs. However, a consistent application of this device would result in musical monotony. Hence the phrases, in particular those of the questioner, are varied. Taken as a whole, this dialogue is structurally unified by its rondo-like form, yet shows diversity through its inner contrasts.

Elements of both these works are contained in the Assumption dialogue 'O superi', set for two sopranos by Carlo Donato Cossoni (1665). Its text is very akin to that of Cozzolani's piece in honour of St Catherine. On the other hand, the addition of a ritornello occurring after each question-cum-answer brings the formal design close to that of 'Psallite superi'. The temporal distance between this work and the two dialogues by the Milanese nun is clearly apparent in the use of explicit contrasts of tempo and dynamics. Moreover, the piece contains very few fragments written in semi-recitative; the human being and the angel address each other mostly in aria style. Although not unattractive, the composition suffers from a weakness also found in other items of Cossoni's first opus: the all too frequent repetition of short-winded phrases on different degrees of the scale. Although this stylistic device belonged to the current practice of Cossoni's time, the composer applied it without the necessary moderation.

Giuseppe Allevi's dialogue dealing with the Assumption, 'Huc omnes currite' (1668), features a 'vox terrestris' (A) and two 'voces caelestes' (SS). This mediocre work shows expert but rather uninspired melodic writing. Moreover, the abundant use of parallel thirds in the parts of the angels produces a sonorous effect that soon becomes tiring. The only interesting aspects of this routine-like composition are the application of the Piacenza formula (see Chapter 1) and the contrapuntal treatment of the voices in the concluding section.

Another Assumption dialogue, 'Quae est ista' by the obscure Roman composer Antonello Filitrani, was set for four parts (SATB) and published in an anthology dating from 1643. Yet its style points to an earlier date of composition, and the layout also evinces an old-fashioned trait: the questions and answers are not consistently assigned to specific voices. This short piece represents the dignified and reserved approach to sacred subjects prevalent in Rome; like so many other works in the early dialogue repertory, it offers its most attractive music in the contrapuntally treated *conclusio*.

More spectacular is Maurizio Cazzati's Marian dialogue 'Silentium omnes' (1668). Its 'cast' consists of personifications of Paradiso (S), Mondo (T), and Purgatorio (B). These allegorical characters do not address one another; instead, they make statements about the Virgin and address the congregation. So this composition could be termed a non-conversational dialogue. Each of the soloists starts with an introduction consisting of a fragment in recitative style followed by a lively short aria. The opening phrase of Purgatorio, with its extremely slow harmonic rhythm, is particularly striking (Ex. 51). Among the noteworthy sections of the composition are two terzets, one with isolated exclamations of 'O' in all three voices, the other provided with the description 'Aria' (which indicates that the term signifies a style rather than a solo piece). Despite the presence of some arresting passages, the mosaic-like structure of the work proves rather unsatisfactory. Cazzati did not achieve the formal unity of the dialogues by Cozzolani and Cossoni.

The texts of two works in praise of unspecified saints, set by Cossoni and Molli, are packed with generalities and clichés; so it is no wonder that they could not inspire the composers. Cossoni's 'Musa voces' (SB; 1665) contains a few

Ex. 51

interesting passages, showing canonic treatment of melodic material, but in other respects it is a routine-like composition. The verbal exchange between the two roles is limited to only fifty-five bars, that is, a quarter of the total length. So, despite its subtitle 'dialogo', the work must be considered a motet rather than a dialogue. 'Beatus vir qui inventus est' by Antonio Molli (1638) is set for four parts (SATB). The repeated initial question, sung by the soprano, serves as a refrain; the other voices in turn provide different answers. The concluding section a 4 cannot redeem this undistinguished composition.

'Quis est iste' by Carlo Grossi—no relative of Giovanni Antonio—is a dialogue between an Angelo (A) and a Huomo (T), written in praise of St Peter (18 January), and dated 1657. Its text offers more interest than those of most other works in this sub-category. The questioner wonders whether it is right to venerate a man who is known to have denied the Lord: 'How can he be called the imitator of Christ?' The angel then explains that Peter deeply regretted his denial, shedding bitter tears, and that, like his Master, he was eventually crucified as a martyr. In this work Grossi presents himself as a capable composer who occasionally achieves striking effects, such as the entry of the angel's voice on the ninth above the bass (Ex. 52). A few instances of word-painting, audible and inaudible, follow tradition: the lowered third on 'dulces lacrimae' and the augmentation of the note *g'*, involving a 'cruciform' sharp as illustration of the word 'crucifixus'.

Ex. 52

Gasparo Casati's dialogue of 1640 in honour of S. Gaudenzio, 'Quam laetam hodie videbo' (22 January), has been mentioned in Chapter 1. Instead of an angel, it is a citizen of Novara who is here questioned by a stranger. The composer rendered the conversation almost entirely in semi-recitative style, the resulting monotony of which is happily interrupted by two attractive triple-time sections.

Totally different is the scoring of 'Quis, putas, puer iste erit?' by Giovanni Battista Fergusio (1612). This is a dialogue about John the Baptist, intended for the feast of his nativity (24 June). The repeated question 'What manner of child shall this be?' (Luke 1: 66) is assigned to a 'canto separato' supported by a theorbo; the answers, full of biblical quotations, are sung by an eight-part ensemble with organ accompaniment. The homophonic setting of the words 'vox clamantis in deserto' is particularly impressive; equally remarkable is the melismatic treatment of the parts of the solo soprano and first tenor (the latter quoting Christ's words: 'Ecce ego mitto Angelum meum ante faciam tuam' (see Matt. 11: 10, Mark 1: 2, and Luke 7: 27)). The work concludes with the text of a Vesper antiphon, 'Inter natos mulierum non surrexit maior Joanne Baptista', set for double choir and combined with a quotation from the Litany of Saints, sung by the solo soprano (Ex. 53). This composition is not only a remarkably early example of a hagiographic dialogue; it is also one of the most moving pieces of Fergusio's 1612 collection. The Turin composer certainly paid a worthy tribute to his name-saint.

Two dialogues by Carlo Milanuzzi, praising St Stephen (26 December) and St Charles Borromeo (4 November), are of particular interest because they disclose their exact (semi-)liturgical function.[1] In the 1636 print both works are indicated as 'Introductio ad Vesperas'. But even without this piece of information the words clearly announce the Vesper service. Written as dialogues between God (B) and a human being (T), they teem with quotations from the Book of Psalms and the Magnificat. Here is the complete texts of the piece in honour of St Stephen (quoted scriptural words in italics):

HOMO. *Exaudi, Domine,*[a] praeces servi tui et *vocem meam,*[a] exaudi in solemnitate sancti Stephani. Et dum te glorificat cor meum, et dum te *magnificat anima mea.*[b] *Memento Domine David et omnis mansuetudinis eius.*[c]

DEUS. Dirige, fili mi, vocem tuam ad me, observa, fili, mandata mea, clama!

HOMO. *De profundis clamavi ad te, Domine. Exaudi vocem meam.*[d] A mandatis enim tuis timui.

DEUS. *Beatus vir qui timet Dominum: in mandatis eius volet nimis.*[e] Timuit Stephanus, servus meus, et confessus est nomen meum. Ideo triumphat in caelis coronatus. *Bonum est*[f] enim *confiteri Domino et psallere nomini*[f] suo altissimo.

HOMO. *Confitebor tibi, Domine, in toto corde meo:*[g] *Sumite psalmum et date tympanum.*[h] Laudate Deum, *laudate in cordis et organo.*[i] Quoniam sic *dixit Dominus.*[j] *Bonum est confiteri Domino et psallere nomini*[f] suo altissimo. Cantate ergo et *psalmum dicite nomini eius.*[k]

[a] Vesper psalm 129: 2.
[b] Magnificat: 1 (= Luke 1: 46).
[c] Vesper psalm 131: 1.
[d] Vesper psalm 129: 1–2.
[e] Vesper psalm 111: 1.
[f] Psalm 91: 2.

[1] C. Milanuzzi, 'Exaudi, Domine, praeces servi tui' (per S. Stefano) and 'Deus, qui vides corda hominum' (per S. Carlo).

Ex. 53

^g Vesper psalm 110: 1.
^h Psalm 80: 3.
ⁱ Psalm 150: 4.
^j Vesper psalm 109: 1.
^k Psalm 65: 2.

It is hardly surprising that this artificial concoction of psalm verses could not inspire Milanuzzi to great music. Yet the tone of both dialogues is dignified and, in addition, occasionally enlivened by the use of melismas and short triple-time passages underlining words like 'cantate' and 'psallite'. The fact that the traditional concluding section set for two voices is missing may be explained by the introductory character of the works. Obviously the first liturgical item of Vespers, the verse 'Deus in adjutorium' and the responsory 'Domine ad adjuvandum', were supposed to be intoned immediately after the dialogue.

Most of the works belonging to the second sub-category oppose a man in distress to an angel who urges him to join the celebration of a saint. Both the celestial and terrestrial voices can be doubled or tripled; in that case they act as homogeneous groups. The contrast between earthly suffering and heavenly joy imparts to these pieces a certain tension. Moreover, unlike the songs of praise, they show psychological development; the sinner or sufferer becomes gradually convinced that, instead of lamenting his lot, he should address himself to the saint in heaven.

Several composers of eulogistic dialogues are also represented in this group of works. It is conspicuous that their contributions are of higher value; this is particularly true of Giuseppe Allevi and Carlo Donato Cossoni. Obviously the dramatic qualities of the subject-matter offered more interest than the artificially dialogued texts of the first sub-category.

Allevi's 'Quam mihi obscura' (1668) is a case in point. Its text mentions only an anonymous male saint: 'Ad sanctum N.—sint tuae preces'. These words are sung by an Angelo Custode (A), who offers consolation and hope to a sinner (B). The latter opens the dialogue with pathetic words in semi-recitative style, emphasizing the antithesis of two different aspects of the day: the feast celebrating the saint reflects clarity, while he himself dwells in darkness. Here as elsewhere in the composition Allevi shows uncommon care over details. His musical prosody is subtle and refined, not least because of the application of the Piacenza formula; moreover, conventional cadences receive a personal touch by means of melodic–harmonic alterations. The desperate utterances of the sinner and the encouraging words of the angel succeed each other at a relatively short time-interval, a procedure that enhances the dramatic impact. When the man finally realizes that salvation is possible, the resulting loss of tension calls for compensation; hence the vocal virtuosity in the final duet describing the joys of eternal life in heaven.

The text of Allevi's 'Fili hominum' (1662), likewise written for the celebration of an unspecified saint, is less appealing. Only the opening phrase of the angel alludes to the vanity of human values; otherwise the piece deals with praise rather than conflict. Since the words do not go beyond generalities, the composer was compelled to write rather abstract music. This is particularly apparent in the complex concluding duet (SA), which occupies about three-fifths of the composition.

A third dialogue by the same composer was written for an unspecified Marian feast. This is 'Infelix anima' (1668), featuring once again the Angelo Custode (S) and a human soul (A). The opening section, written partly in prose, partly in rhymed verses, has already been quoted in Chapter 1; its setting as a semi-recitative followed by a miniature aria is the logical result of the textual layout. During the remainder of the dialogue proper semi-recitative alternates continuously with subsections in triple time. Surprisingly, one of the latter is set to the desperate exclamations of the Anima: 'Heu me miseram, infelicissimam!' These words are underlined by a descending diminished fourth; further on, the same interval stresses in ascending motion the concepts 'poenae' and 'tormenta'. In the *conclusio* both the guardian angel and the human soul express their veneration of the Holy Virgin.

'Ad lacrimas oculi' by Carlo Donato Cossoni (1665) was written 'per qualsivoglia santo o santa'. Yet the text offers the choice between two saints: Antonio, probably St Anthony of Padua (13 June), and St Cecilia (22 November). The prosodic similarity of both names—four identically scanned syllables—is certainly not fortuitous.

Instead of being hampered by the indifferent text, the composer turned its simplicity to good account. A human being (B) complaining about the 'terra deserta et invia' is opposed to two angels (SS) who praise the day of joy. Extremes of tempo reinforce the antithesis: the positive statements are sung presto, the negative ones largo or adagio. One of the latter leads to a curious example of word-painting: an allusion to Psalm 136 ('Hang thy instruments upon the willows') is translated into music by a rest of one-and-a-half bars in the continuo part. Equally noticeable is the descending leap of a ninth on the word 'luctum' (sorrow) (Ex. 54). Apart from this fragment and a few other bars the whole composition is set in triple time with ample use of melismas on words like 'exultandum', 'cantabitis', 'psallite', and 'laetus'. In this way its message, triumph of joy over sorrow, is conveyed effectively to the listener.

Another dialogue by Cossoni, 'Putruerunt et corruptae sunt cicatrices meae', is likewise included in his first opus. The work was written in honour of the martyrs SS. Cosma e Damiano, two brothers who excelled in the art of healing (celebrated on 27, today 26 September). Yet in this case, too, the text offers an alternative subject, namely S. Mauro. Although there are some fifteen saints bearing that name, only Mauro of Glanfeuil (15 January) can be considered here. As a follower of St Benedict at Montecassino, he healed a boy from deaf-mutism. The

Ex. 54

substitution of Mauro for Cosma and Damiano involved some careful rewriting of the text, the more so as the Benedictine prior was no martyr. This is exemplified in the following phrases printed below the music:

Cerne   Cosmam, cerne,   cerne   Damianum: ad   istos   confuge   caelestes
        Maurum, Maurum dantem   medicamen: istum   caelestem

medicos,   in   istos   aspice   martires   inclitos   qui   sanent   te, qui   curent   te.
medicum,   istum   praesulem   inclitum   sanet   curet

O par nobile   martirum   Cosma   et   Damiane!   Eia,   eia   sacro
O lux inclita   caelitum   Maure,   spes   afflictorum!   dulcis,   eia

vestro   cruore,   eia   facite   cordi nostro medicinam.
clemens   dignare,   facere

The dialogue opens with an extensive solo for the Anima (B), who complains bitterly about his wounds (a metaphor for his sins). This is followed by the angel's advice to address the saint(s) in question. During the rest of the work, too, the text adheres to the usual layout found in pieces of this kind. Cossoni's setting is even more brilliant than that of the previously discussed composition, though it is structurally less balanced.

In her Marian dialogue 'Quid, miseri, faciamus' (SATB, 1650) Chiara Margarita Cozzolani once more displays a remarkable capacity to employ current stylistic procedures to their greatest advantage. The distribution of roles is rather unusual. Instead of an angel it is the Holy Virgin herself who shows three sinners the way to salvation. Throughout the work the lower voices sing homophonically as well as contrapuntally in alternation with the celestial soprano. However, in a few instances the group of singers and the solo voice share musical material. The Virgin's text alluding to the Magnificat, 'Invocabitis unigenitum Dei filium, quem ego genui, qui fecit potens, qui fecit mihi magna,' is quoted shortly afterwards by the vocal trio, the original melody being elaborated with ingenious motivic imitations (the words 'invocabitis', 'ego', and 'mihi' are of course replaced by 'invocabimus', 'tu', and 'tibi'). Sister Cozzolani's melodic writing is as simple as it is effective. Ex. 55 shows how the tension, gradually built up through juxtaposition of small varied motifs in ascending motion, finally dissolves in a melisma on the word 'glorificabit'. The phrase is not just a 'correct' setting of the words; the music strongly enhances its communicative function. Speaking of her son, the Virgin tries to convince the sufferers of their possible salvation. As the Virgin can hardly be expected to praise herself, the customary eulogistic conclusion is replaced by a four-part 'Alleluia'. Its metrical ambivalence (6/4 alternating with 3/2) reminds us of Orfeo's 'Vi ricordo, o boschi ombrosi' in the second act of Monteverdi's opera. In this ensemble homophony again appears side by side with counterpoint. Among the composer's important contributions to the dialogue repertory this work is perhaps her most accomplished and effective achievement.

Ex. 55

Any classification of 'Ad arma, gigantes!' (1692) by Giovanni Battista Bassani encounters difficulties. Although termed 'dialogo per la Beata Vergine', the piece hardly belongs to the present sub-category, yet fits still less easily into the other two. Written for soprano and bass, it opposes a Christian to a Turk. This seems to refer to the contemporary wars between the Venetian Republic and the Ottoman Empire; in fact the work has political or—from today's point of view—historical implications.

As a dialogue the composition exemplifies the latest phase of the genre. Containing several items built on prefabricated schemes, it verges on the Latin cantata. The Turk's ferocious opening aria adopts the formal layout A–B–A–C–A.

The Islamic warrior calls his fellows to arms with a melody or rather an elaborated signal consisting exclusively of triadic notes; the text continuously repeats the words 'ad arma, gigantes'. Section B is more explicit: 'In tuba sonora consurgat Aurora et contra Meridiem pugnate.' However, it is only in section C that the intention of this combatant becomes entirely clear: 'Sub vexilo magni Ducis cadat Roma purpurata, et sub Luna coronata pereat. Cadat signum Crucis.' The Christian identifies his enemy with the devil threatening the faith of the occidental world (Recitative: 'Lucifer transibit in Hesperum'). In his ensuing bipartite aria (A–B) he challenges in turn his Muslim adversary, invoking the Virgin for the first time: 'Virgo calcabit superbiam Draconis.' Immediately afterwards, the battle takes place; it is portrayed in a concluding duet with rapid exchange of words (Ex. 56). The formal layout of this duet, identical with that of the Turk's aria (A–B–A–C–A), is quite unsuited to a scene dealing with violent action. The two restatements of section A suggest that, in spite of being vanquished, the Musulman renews his onslaught. Moreover, the text itself contains a fallacy. The words 'Virgo pia, Virgo Maria', spoken by the Christian, are repeatedly interrupted by the Turk asking 'Quae est ista?' While this implies the latter's ignorance about Christianity, rather the opposite is true in historical fact. In the Koran the Virgin (being the only female in this holy book called by her name, Maryam) is mentioned several times and with high praise, notably in the suras III: 42 ff. and 75; XXI: 9; LXVI: 12.[2]

Unlike that of other Marian dialogues, the subject-matter of this particular piece leaves no room for conversion and salvation. Since it represents political actuality rather than legend, an apotheosizing solution of the conflict, as found in Tasso's battle of Tancred and Clorinda, was simply out of the question. From a purely musical point of view the piece is exciting but not really of high quality. Rather, it must be considered a curiosity in the dialogue repertory.

The works belonging to the last sub-category have some resemblance to scriptural dialogues in so far as they portray a scene from the life of a saint. In most cases this concerns his or her martyrdom, a subject involving fierce conflict or even violence. Several pieces deal with an anonymous martyr; these may have been intended for performance on a feast of variable date within the class of the Common of Saints.

Unlike the previously discussed works, the truly hagiographic dialogues show us the saint in action. In addition to verbal exchanges with an adversary, we encounter instances of conversation between two saints, one on earth, the other in heaven. As for the provenance of the texts, the main source must have been the *Legenda aurea* by Jacopo da Varagine (d. 1398). Despite the fact that this famous book was strongly attacked by seventeenth-century rationalists, it was

---

[2] Arabic numbering of verses, adopted in the French edition by D. Masson (Paris, 1967).

Ex. 56

still widely read and admired; this is shown by quite a few literal quotations in the works to be discussed below.

Practically all the dialogues concerning martyrdom were composed after 1650. Although several of them are preserved in undated manuscripts, the musical style leaves no doubt about this. The sole exception is 'Viri timorati' by Giovanni Ceresini, who at the time of its publication (1617) directed the chapel of the Accademia della Morte in Ferrara. This piece contains a conversation between the protomartyr St Stephen (S) and four adherents of Christ, called 'fratres' in the text. The latter, expressing fear of their impending persecution, are mentally fortified by the saint: 'Lapides torrentes mihi dulces fuerunt propter Iesum stantem a dextris virtutis Dei in auxilium mihi et populo sancto eius.' The concluding ensemble a 5 is a song of praise; in addition, it expresses confidence in divine assistance. The musical setting is dignified, yet rather neutral. The composer makes no attempt to depict either the feelings of the protagonist or those of the group.

Monferrato's 'Mea lumina, versate lacrymas' (1681) shows how much more expressively a similar theme was treated some sixty years later. This dialogue was written in honour of St Lawrence (10 August); apart from the saint (T), it features two angels (SS). The first of these opens the work with a moving complaint about the torture undergone by Lawrence. In the following section the Roman archdeacon himself calls upon Jesus, in whose presence he no longer fears excruciation: 'laetus imbres sagittae . . . sustinebo'. Both these solos are built on

musical phrases with recurrent motifs. A restatement of the opening words is set
for three voices. The tenor's participation in this short homophonic ensemble is,
strictly speaking, anomalous, yet not really disturbing, as the voice merely dou-
bles the continuo. In the subsequent duet the angels cry out to the torturers,
asking them in vain to stop. The dying Lawrence then commends his spirit to
Jesus's hands. The *conclusio* starts with a lively duet of the celestial voices,
singing of joy and triumph; after twenty bars they are joined by the saint, who
exalts his spiritual victory.

An even stronger musical expression of the words is found in 'Heu me!
miserum me!' by Giovanni Antonio Grossi. In the opening section (see Ex. 7 on
p. 23) the agony of an unspecified martyr is depicted with an intensity which
involves entries on unprepared dissonant notes, such as a minor seventh (bar 4)
or even a diminished octave (bar 7). Equally noteworthy are the harmonic dissonances within the phrases: A♭–F♯ in bar 5; B♮–B♭ in bar 14; F♯–F♮ in bar 20. The
martyr does not call upon heaven; instead, he himself is addressed by
S. Liborio: 'Euge, euge, cur infelix quaeritas . . . cur infelix quaerulas? Remedia
quaere, opem pete et petrarum onere liber eris.' After several repeats, of both
the lamentation and the consoling words, the sufferer asks to whom he is
indebted for this assistance, whereupon Liberio identifies himself and achieves
the miracle, grinding the stones to powder. During the remainder of the dialogue the saint is praised by the martyr, who in this case has escaped from death.
This explains why the work was written in commemoration not of him but of
S. Liborio (whose feast is on 23 July).

Among the posthumously printed compositions of Bonifazio Graziani are two
dialogues representing anonymous martyrs, one female, the other male. 'Infelix,
non vides' is a scene opposing a Christian virgin to a heathen tyrant who tries to
seduce her. The formal disposition of the first part of the piece illustrates the
composer's clever handling of the text. The tyrant opens with a small aria,
threatening the virgin (section A) but also blandishing her (section B). One
would expect a restatement of the first section, rounding off the miniature da
capo form, but this does not happen. Instead, the Vergine Santa sings an even
briefer aria, telling the tyrant that she is neither impressed by his intimidations,
nor taken in by his flattery. Although this arietta is based on a tripartite scheme,
its da capo section is extended by five extra bars. The concise dimensions are
shown in the following table:

Form:	A		B		A
Keys:	A minor → E minor		C major → A minor		A minor → E minor
Bars:	4		11		9

The mounting tension calls for a duet sung in semi-recitative style. Its textual
phrases are effectively divided between the two characters, entailing a gradually
accelerating exchange of words:

TYRANT

Ergo tuam vis perdere iuventutem . . .
Si despicis venustatem . . .
Sed si ferrum nudabitur . . .
Si flamma parabitur . . .
Unde virtus?
Quis se firmat?

VIRGIN

ut merear aeternitatem.
ut vultum cum sole commutem.
collum curabo.
ignem intrabo.
Ab excelso.
Fides Christi.

It is only after this lively duet that the tyrant restates the first section of his opening aria; in this new position, it sounds all the more threatening. The virgin repeats her refusal in stronger words than before ('Perfide, barbare, seducere tentas me'), singing a second da capo aria. Her fortitude and determination are underlined by the note *g"* sustained over four bars ('immobilis sto') and ascending scales ('sum fortis in spe'). Now the tyrant abandons his attempts to seduce the girl. Working himself up into a frightful rage, he evokes all the destructive forces of nature and hell. Since this outburst does not lend itself to a regular formal scheme, his aria—again of very modest proportions—is through-composed. The dialogue proper ends with a likewise unschematically organized duet which opposes the virgin's absolute confidence in God to the bloodthirsty commands given by the tyrant ('Rapite, frangite, scindite membra'). In the *conclusio* the two characters step outside their roles and address the listener: 'Videte, quam prompta, quam laete suscipitur pœna.'

The other dialogue, 'Ecce furor', was written for the celebration of a male martyr. Its scoring for three sopranos is exceptional. In view of the high tessitura of the parts, two of which ascend to *a"* and one to *b"*, they can hardly have been meant for boys. In all probability the work was performed by castratos or (less likely) falsettists.

Three Christians feel themselves to be in acute danger, threatened as they are by the heathens. They express their fear in a short terzet preceded by a recitative. Two of them urge flight: 'Fuge, Christi miles, fuge impios hostes, fuge mortem.' The third, however, chooses a martyr's death and earns eternal life. He announces his decision in a semi-recitative followed by an arietta. His companions comment on this choice in a delightful little duet ('Vade felix, vade laetus'). From this point the dramatic development comes to a standstill; firmness on the one side and sympathy on the other are repeatedly expressed in solos and duets. The work concludes with the usual address to the congregation, sung by all the three voices. Although musically attractive, the dialogue suffers from a lack of dramatic tension on account of the rather tedious text.

Two works preserved in manuscript in the Cathedral Archives of Como were certainly composed in that town. Each contains only a single scene, opposing once again a martyr to a tyrant. The anonymous dialogue 'Age furor, urge minas', set for three voices, takes us *in medias res*. The tyrant (B) orders the Christian (A) to be tortured. The latter, fortified by an angel (S), defies him. There are practically no separate solo sections in this piece; instead, the composer recounts the contest in a rapid exchange of words (Ex. 57). This brief through-composed work dispenses with a moralizing *conclusio*. The antithesis is maintained until the end; the angel and martyr exalt their triumph, the tyrant persists in his fierce language (S and A: 'vives/vivam in aeternum'; B: 'morieris in aeternum').

Ex. 57

Whereas in this dialogue violence is expressed by purely vocal means, other pieces achieve the same effect through the addition of instruments. The second Como work, 'Cede Constantia' by the local composer Francesco Rusca, is set for five voices (SSSSB) and two violins. Its unbalanced scoring, showing an unsatisfactory preponderance of high vocal and instrumental parts, has been mentioned already in Chapter 1. A performance with transposition of the three assessors' parts to the lower octave is possible, however. This might even have been the composer's original intention.

The dialogue was written in honour of the almost totally unknown Santa Costanza, who seems to have been venerated exclusively in the Como region and southern Ticino (celebration on 24 February). Its text is broadly comparable to that of the anonymous piece. The assessors reinforce the part of the tyrant; even when singing without him, they repeat his words. The violins, which open the work with a few bars written in *stile concitato*, have a double function. They set the tone of the men's violence but provide in addition a halo for the saint; Ex. 58 shows how she brings out the literal meaning of her name. In this attractive piece the 'actors' maintain their identity until the end, leaving the tragic outcome to the listener's imagination.

Ex. 58

The style of Ippolito Ghezzi's dialogue 'Tacete tyranni!' (SB, 2 violins, 1708) closely resembles that of his 'Bella, pugna', the battle between St Michael and Lucifer, discussed in the previous chapter. An anonymous male or female martyr and a tyrant, each representing a group, are involved in a fierce dispute, the violent result of which seems inevitable. Unlike the two Como dialogues, this piece includes three items written in da capo form: one duet and two arias. During the duet the violins, playing in *stile concertato*, stress the vehemence of the conflict; the same effect is achieved through sequences of descending scales in the tyrant's aria. At this point the action takes an unexpected turn. In a short recitative the Christian prays to God that he may open the eyes of his adversary and in the subsequent aria (set to verses) he tries to convince the tyrant:

> Si, tyranne, Christo credis
> et ad Iesum corde accedis,
> mihi erit laeta dies,
> tibi erit vera quies
> et arridet fausta sors.

This most expressive 'aria affettuosa', accompanied by the violins playing in unison, is full of Neapolitan sixth chords and dynamic contrasts. No wonder that it achieves its effect. The tyrant stands confounded: 'Quis affligit sic cor meum? Morior, cado. Heu me!' The Christian repeats his urging, assuring him that he will not die. Then the tyrant yields ('Spero, credo, adoro') and the work concludes with a joyful duet in 6/8 time.

Three hagiographic dialogues dating from the first half of the century deal with less violent episodes. 'Dixit Laurentius' (1629) by Ignazio Donati represents a conversation between Pope Sixtus II and St Lawrence. The scene's rather complicated antecedents, related in the *Legenda aurea*, are the following. Brought from Spain to Rome, Lawrence became the pope's archdeacon. At that time—the third century—the Emperor Philippus, a Christian, was forced by the rebellious general Decius to leave the city. Before his departure he entrusted Sixtus and Lawrence with the imperial treasure, telling them that, in the case of his defeat, it should become the property of the Church and the poor. The rebel, having seized the throne, began to persecute the Christians; among the numerous victims was Philippus himself. Then Decius summoned the pope and ordered him to deliver up the treasure. Threatened by force, Sixtus had to admit that it was now in the Church's possession. This admission bewildered the archdeacon, and it is at this point that the dialogue starts. Apart from Lawrence (S) and Sixtus (B), there is an important role for the narrator, whose words are divided between the two middle voices (AT). It was probably for this reason that Donati headed his work with the caption 'in modo di dialogo'.

The opening phrase, sung by the Historicus, may seem puzzling: 'Dixit Laurentius ad Beatum Christum'. Yet the saint does not address Jesus. The last two words should be read as 'blessed anointed', that is, the pope. Lawrence's question, 'Quo progrederis sine filio, Pater?', is a literal quotation from the *Legenda aurea* and so is Sixtus's answer, 'Non ego te defero.' The pope tells him then to distribute part of the treasure to the churches and the poor. However, over the following days the archdeacon gives the whole of it to the poor and asks the pope's forgiveness for this. Sixtus states once more that he (Lawrence) shall not be called to account, adding, however, that the struggle for the sake of Christian faith awaiting him will be greater than his own (another quotation from Jacopo). No explicit mention is made of the saint's subsequent martyrdom, but the dialogue ends with his prayer: 'Domine Jesu Christe, miserere mihi, servo tuo.' The five-part *conclusio* addresses the congregation with praise of God and his saints. Although the work is not outstanding among Donati's dialogues, it shows the characteristics of the composer's great abilities: beautiful melodic writing with rhythmic subtleties and masterly handling of counterpoint, in particular in the concluding section.

Unlike this work, 'O Catinensis gloria' by Michele Malerba (1614) is an insignificant composition. Yet its subject and presentation are interesting. The piece deals with two female saints, Sant'Agata (in heaven) and Santa Lucia (on earth). Like the composer, both were Sicilians. In this case, too, the events preceding the scene need to be retold. Lucia, a noble virgin, lives with her mother in Syracuse. The latter has been suffering for four years from a haemorrhage, a disorder that seems incurable. It is for this reason that mother and daughter go to Catania, the city of Sant'Agata. During a church service they hear that Jesus healed a woman suffering from the same disease (Matt. 9: 20–2; Mark 5: 25–34; Luke 8: 43–8). This leads Lucia to pray at Sant'Agata's tomb. In a vision she sees the saint, crowned with jewels and surrounded by angels. She then begs Sant'Agata to cure her mother, but the saint replies: 'Why, sister, my Lucia, are you asking of me that which you are able to perform yourself? Behold, it is through your faith that your mother is healed.' Out of this story the short dialogue picks up only Lucia's question and Agata's answer, both freely borrowed from the *Legenda aurea*. The conclusion a 2 (SA) glorifies the saint of Catania, calling her 'Regina Trinachiae' (Queen of Sicily). This indicates that the work was written for the celebration of Sant'Agata (5 February).

Finally there is a pleasant, almost naïve, dialogue between Jesus (A) and Mary (S) on the occasion of the latter's assumption in heaven, 'Ave Mater dilectissima' (1642). Its composer, Chiara Margarita Cozzolani, was the right person to render this text with convincing music. Nothing much happens in this scene. Jesus tells his mother that the days of sorrow are over; he and the saints welcome her most warmly. The Virgin in turn expresses her delight at hearing her son's sweet voice. She is happy to die for she now sees him in his triumph. In the course of the composition textual phrases recur with slight variants. Also, a few melodic fragments are repeated in both vocal parts; this procedure not only reinforces the community of feelings but in addition enhances the unity of musical structure. Among the most attractive details is the concluding syncopated 'Alleluia', occupying only nineteen bars. The work is a characteristic example of a type of dialogue which combines ingenuous words with no less ingenuous, yet masterfully conceived, music.

Moralizing Dialogues

Allevi's 'Peccatrix anima' (1662) shows us an angel giving a severe warning to a soul:

In sordibus mundi immunda dormis; onusta maculis, inhonesta vitiis tui oblita Creatoris et amoris es. Tu meas audi voces et inhorresce, perfida: peccata damnant te et poenae manent te.

(Impure as you are, you sleep in the foulness of the world; tainted with disgrace, degraded by crimes, you are oblivious of your Creator and charity. Listen to my words, you infidel, and beware: your sins condemn you and punishment is in store for you.)

Deeply touched by these words, the soul expresses her fear of eternal damnation. The angel, however, holds out a prospect of salvation on condition that the sinner undergoes conversion. This happens, and so the piece concludes with a joyful duet referring to the redeeming blood of Jesus: 'te/me cruor Salvatoris sanabit, purgavit'.

This moralizing text, which opposes heaven to earth, exemplifies a type of dialogue that dominated the repertory in the middle of the century. Variants in the distribution of roles are, of course, encountered. The angel is sometimes replaced by God or Jesus. As for the soul, she is not invariably cast as a sinner but also appears as a being utterly dissatisfied with the *vanitas* of a worldly existence. Although in general there are only two characters, works with additional roles, for instance a group of sinners or a narrator, are also found. A fairly constant characteristic of the music is the predomination of semi-recitative style. Obviously the aria was considered as unsuitable for the setting of a text containing an argument. Therefore in the pieces adhering to this principle only the concluding section offered the composer the opportunity to apply musical procedures other than expressive monodic writing.

In view of the great number of moralizing dialogues, especially in the northern regions of Italy, there must have been a need for these 'didactic' compositions. Pieces dating from the first three decades of the seventeenth century are rare, however. Banchieri's *Cuor contrito al suo Creatore* (1625) opposes a sinner (S or T) to God (B). The concise text, typical of the early phase of the genre, does not leave much room for the musical expression of emotions. Yet the rests separating the words of the sinner are conspicuous ('Hei mihi, Domine, vide . . . humilitatem . . . meam'). As is usual in Banchieri's 1625 volume, one of the interlocutors (in this case the Creatore) loses his identity towards the end of the piece, the text of the other being sung by both voices.

'O mi Domine' by Chiara Margarita Cozzolani (1642) is almost four times as long as Banchieri's dialogue. A repentant soul (A) addresses a guardian angel (T), asking for aid. The angel's advice is predictable: 'Clama ad Dominum, confitere, ingemisce, suspiria emitte, lacrymas effunde.' The conversation, which

Ex. 59

spreads over more than two hundred bars, contains a few remarkable details, for instance the deceptive cadence underlining the word 'iniquitates' and leading to a modulation (Ex. 59). In spite of this, Sister Cozzolani could not entirely avoid the risk of tediousness due to the continuous use of semi-recitative. This weakness is apparent in quite a few moralizing dialogues of the time.

Shorter pieces by other Milanese composers are more satisfactory. Although 'Ubi es, o mi care Iesu', set by Michel'Angelo Grancino (1631), is certainly not a distinguished composition, it lacks the monotony of Cozzolani's dialogue and brings a climax in the homophonic *conclusio* with two additional voices. Francesco Bagati's 'Quid clamas, filia?' (1658) opposes a woman to Jesus. The latter's part includes a melisma of quavers and semiquavers lasting seven bars. Set to the word 'desponsatio', this forms a welcome contrast to the predominantly syllabic setting of the rest of the text.

Composers working in Piacenza used to treat moralizing subjects in a fairly conventional way. In addition to 'Peccatrix anima', Giuseppe Allevi wrote the dialogues 'O Anima' (1654) and 'Quo progrederis, dilecte mi' (1662). Neither of these pieces contains anything spectacular. The only noteworthy detail, occurring in the first of the two compositions, is of a textual rather than a musical kind: the insertion of an Italian exclamation in a Latin context (*'Deh* veni, iam veni'). Yet Allevi's settings are not at all tedious. His careful melodic writing with subtle handling of the musical prosody captures the listener's attention. The composer's pupil Isodoro Tortona strictly imitated the style of his master; this is apparent from the dialogues 'Quid sentio' and 'Vulnerata charitate', included in Allevi's collections of 1662 and 1668, respectively. Contrary to what one would expect, this stylistic dependence did not result in loss of quality. 'Quid sentio', in particular, is a work of great beauty, showing a perfect balance of melody, rhythm, harmony, and (in the concluding duet) counterpoint.

The danger of monotony induced several composers to employ procedures enhancing the inner variety of their works. 'Heu dolor', a dialogue by Nicolò Fontei (1638), features a man (A) and Jesus (B). In the course of the piece the latter's words 'ut pugnes adhortabor, ut vinces adiuvabo, et si Deus pro te, quis contra te?' call for martial music. The composer does not let the opportunity slip; both this text fragment and the following ('te certantem spectabo, deficientem sublevabo, vincentem coronabo') are set in genuine *stile concitato*, the voice suggesting the sound of the trumpet, the continuo that of the drum.

A more radical device was employed by Cossoni in his 'Cogitavi dies antiquos' (1670). This dialogue between a soul (S) and God (B) is unexpectedly interrupted at the mention of eternal damnation. Temporarily abandoning their roles, both voices interpret the words 'O terribilis aeternitas, o immutabilis aeternitas' by repeated notes sung in unison, leaving the melodic movement to the instrumental bass. The opening phrase of this dialogue is set to the sixth verse of Psalm 76. Although in general the moralizing texts are freely invented, sometimes the authors borrowed from the Scriptures.

An original composer like Giovanni Antonio Grossi did not submit to convention. In his 'O fortunati dies' (1670) the soul, fond of her worldly life, offers more resistance than in other pieces of this kind. Grossi renders her profanity by a florid melody above a *basso ostinato* (the usual descending tetrachord). The words 'iniqua es' are depicted aptly through an unprepared dissonance, a diminished octave between the voice part and the continuo; the same interval occurs ironically at the moment when the soul boasts of her well-being: 'sum dives, sum pulchra, amata sum'. A duet in the middle of the work, containing a rapid exchange of words, reinforces the antagonism. Then the angel makes a final effort to convert the pleasure-loving soul, opposing the joys of heaven to the torments of hell. This has its effect, and the voices join in a concluding expression of relief: 'falsa valete'.

Tension is also present in two superb dialogues by Giovanni Legrenzi. 'Peccavi nimis in vita mea' (1660) opens with a complaint of a repentant sinner (T) who subsequently confronts God's implacability (B). Instead of expressing their thoughts in long solo phrases, the two parts alternate at a short time-interval; further, they sing simultaneously. The dramatic effectiveness is enhanced by abrupt changes of tempo. It is only after eighty-nine bars that God, having become convinced of the sinner's sincere remorse, grants him forgiveness, and so the piece concludes with expressions of joy on both sides. A noteworthy aspect of the text is the quotation of liturgical and biblical words in the tenor's part, alluding to a responsory from the Office of the Dead,[3] as well as to Psalms 6 and 50.

'Cadite montes' (1655), a dialogue between a man afflicted by poverty (B) and a consoling angel (S), demonstrates similarly Legrenzi's care over a balanced structure. The opening phrase of the bass is a desperate outcry covering his entire range of two octaves (Ex. 60). After a few bars sung in slow tempo ('tot mala me premunt, tot curae me cruciant, tot damna me quassant') the opening phrase is repeated in a shortened version. The gentle voice of the angel provides a striking contrast to this violent outburst. Yet it takes a considerable time for the

---

[3] The quotation is from the responsory 'Hei mihi, Domine' (Matins, 2nd Nocturn).

Ex. 60

man to realize the paradoxical condition of earthly life: 'Omnia mala bona, omnia damna dona.' Subsequently, he indulges in self-reproach: 'O me stultum, insensatum! Comparatus sum iumentis insipientibus et similis factus sum illis.' Assisted by the angel, he then addresses heaven: 'Parce, Domine, et miserere mei/ei.' As there is no role for God or Jesus, the answer is left to the listener's imagination. However, the ensuing short repeat of the dialogue's opening phrase—now in an entirely different context—leaves no doubt about its positive character: 'Cadite montes, percute caelum, degluti terra, numquam me separabis a Redemptore meo.' This repeat is a stroke of genius on the part of the composer. Not only does it eliminate the danger of monotony, but it also serves the unity of the composition as a whole. Although the text of this work does not contain any scriptural quotation, it clearly alludes to the Book of Job.

The decline of semi-recitative style during the last decades of the seventeenth century is apparent in two dialogues by minor masters, Carlo Giuseppe Sanromano and Bartolomeo Trabattone; these date from 1680 and 1682 respectively. 'Cara Maria' by Sanromano consists almost entirely of arias and sections written in aria style. Musically, the piece is not unattractive, but the subject—a prayer addressed to the Holy Virgin—calls for something more personal than the almost mechanical repetition of motifs and cadences on various degrees of the tonal scale. Trabattone's 'Care Iesu' (*Dialogo tra Dio e l'Anima*) is written throughout in triple metre with abundant use of melismas. Only in the introductory section of the concluding duet is there a short fragment in common time. Unlike most moralizing dialogues, the present example reaches no climax. The extensive canonic treatment of uninteresting melodic material in this duet soon becomes boring.

Moralizing texts set for three to five voices offer greater opportunity for musical variety. In 'Quo pergis, amate Iesu' Giovanni Antonio Grossi opposes two souls (AT) to Jesus (B). Each of them has only short solos; elsewhere they sing simultaneously in counterpoint as well as homophony. Moreover, the dialogue proper is interrupted half-way by a short ensemble set for all the three voices ('O cara, o dulcis, o suspirata amantis Jesu vigilantia'). As the rather conventional text could hardly serve as a source of inspiration for the composer, the qualities of the work are purely musical rather than dramatic. Particularly noteworthy are instances of the use of the Piacenza formula in the prosody and the integration of the continuo in the melodic structure of the concluding section. In the course of the piece Grossi widens the ambit of the main key (C minor). Yet, despite cadences on D flat major and B flat minor, there is no question of true modulation in the tonal sense. This work shows us once more the composer's remarkable versatility.

In view of his great reputation in the field of sacred music, Giovanni Paolo Colonna's dialogue 'Adstabat coram sacro altari' (1668) is rather disappointing. The opening phrase is sung by a *testo*, a completely superfluous role that cannot have been added to those of the actors for any other than purely musical reasons. The principal motif of the soul's G major aria (D–B–D–A) is a Guidonian transposition of the initial syllables of the text: '*Sol, mi sol re*surge', words that reappear in her duet with Christ. This may be very clever, yet it does not produce music of high quality. It is only in the concluding 'Alleluia' that Colonna, no longer dependent on an uninspiring text, shows us his true face, writing a delightful little fugue based on two *soggetti*.

Another three-part dialogue belongs to the last phase of the genre. This is 'Vulneratum cor meum' by Francesco Antonio Urio (1690), set to a rather philosophical text featuring a soul (S), a personification of Death (B), and Jesus (A). Today Urio is almost exclusively known as the composer of a Te Deum, the material of which was borrowed and reworked by Handel for use in several oratorios and his own Dettingen Te Deum. The present work testifies to Urio's capacities in the small *concertato*. Unlike the pieces of Sanromano and Trabattone, it opens with a section in semi-recitative involving all the three characters. This makes the ensuing arias of Death and Jesus the more acceptable. Since only about two-thirds of this composition was available for the present study, its final part must be left undiscussed.

Although Protestants could hardly have been offended by the texts of many moralizing dialogues—a fact corroborated by German reprints—they would certainly have objected to that of Grancino's 'Vae mihi misero', because of the mediating role of the Holy Virgin. The work, published in 1631, is scored for five voices. A complaint of a miserable sinner (T) is answered by two angels (SS), who urge him to pray. When he asks to whom he should appeal, they tell him to address his prayer to the Virgin ('Respice Stellam, invoca Mariam, unicam Caeli portam'). Knowing that Mary cannot act for herself, he begs her to intercede in his favour. Subsequently the Virgin (A) appeals in turn to her son with an expressive plea for the sinner's salvation. It is only then that he is forgiven by Jesus (B), whose words include the mention of his mother's intercession. The rather complicated process seems to call for an extensive musical treatment of the text, but this is not the case. The dialogue proper is rather short (a hundred bars) and so too is the concluding five-part ensemble, which addresses the congregation in homophony: 'Iubilantes et exultantes voce sonora cantemus: Iesus est qui culpas et supplicia remittit et dat praemia.' The work is superior to the dialogue 'Ubi es, O mi care Iesu?' by the same composer; not only because of the greater number of roles offering more variety but also because of the succinct and penetrating rendering of the text in music.

Finally, there is a four-part dialogue by Alessandro Grandi: 'Heu mihi!' This work is the earliest moralizing piece (1613); it is also the most impressive in the repertory (see the transcription in Part Two). A repentant sinner (T) addresses three anonymous characters, presumably angels (ATB). However, in the course of the work a 'stage direction' indicates that one of them should make himself invisible and change his identity into that of God, responding in echo to the sinner's prayer. The originality of this dialogue is further demonstrated by various details: the indeterminate chromaticism in the part of the protagonist (bars 14–15), expressing his insecurity, the trills on 'tremor' (bars 18 and 23) in a phrase quoting Psalm 54: 6, and the dissonances in the part-writing (bars 7, 9, 72). Compared with the dark colour of this profound work—the upper voice, the altus, is actually a high tenor—many a piece from the mid-century may seem 'frivolous' indeed. However, it is doubtful whether contemporaries were of the same opinion.

The principal characteristic of the moralizing dialogues discussed so far is the conflict between heaven and earth, personified by an angel, Jesus, or God on the one hand, and a soul or a human being on the other. The result is invariably the latter's conversion and so the confrontation resolves in harmony. Apart from this main model there are two other types belonging to relatively small sub-categories. One of these is the dialogue including a negative character who, consistently keeping to his role, disturbs the final euphoria. The other type presents a dispute in heaven about the fate of the earthly sinner, the outcome of which is his or her salvation. A number of examples chosen from both groups will be discussed here.

Dialogues including a negative character reflect in a simplified way the framework of Christian faith before the age of Enlightenment. The *peccator damnatus* is the necessary complement to the *peccator beatus*: man goes either to hell or to heaven, and there is no other choice. As for the devil, his role, too, is indispensable. Like Goodness, Evil needs its representative in the spiritual cosmos. It goes without saying that the engagement between positive and negative forces, with the possession of the human soul as the stake, considerably reinforces the dramatic quality of dialogues of this kind.

A characteristic example is a work by Giovanni Antonio Grossi, 'Fulcite me flores', featuring an angel (S), a soul (A), and the devil (B). The piece starts with the soul expressing her delight in worldly life: 'mundae blanditiae, luxus ac pompae, stipate me'. The unusual opening melody with its tickling descending fifths (Ex. 61a) is repeated further on almost literally by the angel; set to the words 'perire est' (it's your downfall), it now produces a particularly ironic effect (Ex. 61b). Frightened by the angel's sombre prediction, the soul exclaims: 'O misera! Quo me delector, quid amore prosequor?' At this moment the devil makes his entry; proclaiming the sweetness of his nectar, he tries to seduce the soul. Subsequently there is a long dispute between the angel and the devil. Both of them also address the soul, who remains in a state of confusion and fear. Particularly remarkable is the devil's musical phrase depicting 'the path of Christ, too narrow and full of tears'; this takes the form of an ascending diatonic scale followed by descending chromaticism. Finally, the angel succeeds in winning the soul to his side, and the dialogue concludes with a terzet, written partly in brilliant triple counterpoint and expressing the opposing feelings about the soul's salvation: 'O quantum gaudium/quantus terror super uno peccatore penitentiam agente, o quanta laetitia/tristitia iam sit in caelis/inferis!'

Ex. 61

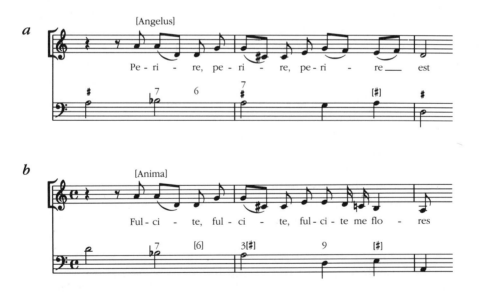

The scoring of 'Fulcite me flores' is symbolic: the highest voice is assigned to the angel and the lowest to Satan, while the soul lies in the middle. This scoring is also found in another dialogue by Grossi ('Quo Domine benignissime'), treating the same subject. Here the three roles are doubled: two angels (SS) are fighting two demons (BB) for the possession of two souls (AT). Each of the three couples sings predominantly in homophony, reducing in this way the six-part scoring to a complex trio setting. The work opens with a phrase of the souls, who show their insecurity by asking: 'Where, O Lord, where, O most gentle Christ, where should we go, what should we do to enjoy the felicity of the blessed?' The angels' answer ('Fear the Lord, love God, follow his precepts, and you will be saved forever, you will be blessed in heaven') is immediately contested by the demons, who offer the souls all sorts of delight: 'shun the spines, take the roses and lilies, and you will be happy forever'. The question and the two statements are each set in different manner, identifying the characters. Easy to seduce, the souls are taken in by the devils' words, but the ensuing severe warning by the angels brings them to their senses, and from here on it is a battle of four against two. The demons are chased back into hell, complaining about their fate with musical word-painting (bars 84 ff.: written-out trills on 'tremor'; bars 118 ff.: suspensions provoking strong dissonances on 'nos lugebimus aeterni, nos plorabimus in saecula'). The piece reminds us of Grossi's *Battaglia di S. Michele Archangelo col Demonio in cielo*, discussed in the previous chapter. Both works achieve their effect by simple means, a characteristic of compositional mastery.

A third work by Grossi, likewise involving demons, is remarkable not only because of its musical qualities but also because of its textual properties. This is 'Currite venatores', a dialogue a 3, opposing two devils disguised as hunters (BB) to an angel (S). There are no parts for the souls; yet they play an important, though passive, role in an allegorical context. Described as dogs, they appear in two capacities, as hounds ('utiles canes') and as quarry ('praeda'). In this way the souls are shown in the double function of robbers and robbed. However, through the power of the angel the demonic hunters meet with the fate intended for their victims. While the celestial voice triumphantly proclaims the 'raptores rapti', the devils concede that they have become 'venatores capti'. The paronomastic character of the text is exemplified by several puns, as, for instance, the hunters' words 'est venia venandi' (it's our privilege to hunt). Musically, the work abounds with symbols of hunting: horn-like motifs, signals set to onomatopoeic words ('taratantara'), and almost continuous canonic treatment of the basses, a traditional procedure reaching back to the fourteenth-century Florentine *caccia*. Like 'Quo Domine benignissime', this composition achieves its spectacular effect by particularly simple means, such as diatonic melodies, elementary chords, and slow harmonic rhythm.

Two other dialogues including a role for the devil pale before Grossi's brilliant settings. Little need be said about 'Quid divitias te caedis' (1658) by Francesco Bagati, a conversation between Satan (B) and a soul (A). The all too frequent repetitions of small musical units stamp this work as a rather tedious composition. Although Gasparo Casati's 'O angele' is a piece of higher quality, the composer sacrificed his freedom of expression to the intelligibility of the verbal message; with the exception of the concluding ensemble, the whole dialogue is strictly set in semi-recitative. The didactic text is indeed interesting and worth quoting in free translation:

DIABOLUS. O Angel, why do you allow human beings to be condemned to hell, why do you not protect them against their sins?

ANGELUS. O malevolent spirit, why do you not content yourself with the fire, rightly prepared for you by the good Lord, instead of dragging man along with you to his doom?

DIABOLUS. This is entirely your fault, because you fail to protect him. We, on the other hand, being damned, cannot suffer man to choose heaven.

ANGELUS. How was it that you, enjoying a blessed life, did the wrong thing and got yourself thrown into hell because of your megalomania? (*Drawing the man into the conversation*) You who have listened to all of this, why do you obey your enemy, the devil, instead of abstaining from all these bad sins?

HOMO. You know that I am totally dedicated to God, yet I am strongly corrupted by the devil. This is no wonder because, as you know, I was born weak.

DIABOLUS. Why do you accuse me, Man, of all these sins? They were not conceived by me. It was you who committed them uncoerced. You joined the damned by your free will.

HOMO. Now I see that you are demented. Whoever follows you in worldly matters deceives himself. Go away, you malign ghost, you are out of your senses.

This text touches upon such fundamental questions as the origin of evil, free will, and personal responsibility. Although the debate does not seem very fruitful, the subject's treatment goes far beyond the simplistic statements made in other moralizing dialogues. As an expert composer Casati was able to set the words in neutral semi-recitative style without falling into monotony. The concluding terzet (presto) offered him the opportunity to show his proficiency in the art of counterpoint.

It seems that for Grossi the unrepenting sinner offered less appeal than Satan himself. His setting of 'O felicitas divina', a conversation between an 'anima penitente' and an 'anima peccatrice', followed by God's statement, lacks the brilliance of the three dialogues discussed above. The reason for this may have been the unsatisfactory text. The words of the celestial voice

> et amicus de hoc mundo
> semper iacet in profundo

seem to point to the sinner's damnation. Yet without any explanation the latter praises Jesus and renounces the world in the concluding terzet.

Unlike this composition Allevi's three-part dialogue 'Iam caeli micant flores' (1668) includes a role for a true *peccator damnatus*. In the final section of the work he deplores his fate: 'Delebo in inferno, ubi nulla est redemptio' (an allusion to a responsory from the Office of the Dead: 'Peccantem me quotidie et non me paenitentem').[4] Yet the text of this dialogue, too, contains a detail that seems

---

[4] Matins, 3rd Nocturn.

rather odd. After 122 bars the angel for no apparent reason mentions an unidentified male saint, indicated by the usual formula N—set to three notes. This was possibly done to make the work available for various ecclesiastical feasts.

Allevi's music shows the characteristics already discussed earlier in this study: expert melodic writing with refined prosody but lack of the originality and versatility found in Grossi's dialogues.

Pieces depicting a dispute in heaven can take the form of a heated discussion between allegorical characters or a difference of opinion about the sinner's fate, opposing in one case the Virgin to Jesus and in another, surprisingly, Jesus to God. An example of each type will be described here.

Natale Monferrato included in his Op. 18 (1681) an allegorical dialogue entitled *Triumphus divinis amoris*. Personifications of two concepts, Iustitia and Misericordia (both S), are engaged in a vigorous dispute. The former urges God in particularly forceful language to punish a sinner: 'Why do you hesitate to take revenge? Punish the crimes, hurl your lightnings, let the world perish!' (the last-quoted words show that the anonymous sinner represents mankind in general). Misericordia strongly opposes this view, furiously addressing her adversary ('Cessa, oratrix importuna!') and calling for grace instead of arms. The dispute continues for a considerable time. Cries like 'Ad arma, ad bellum, ad pugnam!' on the one side are countered with the words 'Tace!' and 'Obmutesce!' on the other. Finally God (B) interposes: 'Silentium! Ego Deus, ego Pater. Nesciunt paterna viscera vindictam. Misereor orbis . . . Ignosco culpis. Convertatur homo et vivat.' Subsequently a concluding ensemble a 3 rejoices at the salvation of mankind.

Dialogues by Monferrato discussed in Chapter 3 and earlier in this chapter showed the composer's ability to render tender or elegiac feelings. Here he reveals his dramatic talent, reinforcing the martial language of Iustitia with trumpet and drum motifs in the vocal part as well as the continuo (both instruments are mentioned in the text). There is frequent change of tempo, and the parts of Deus and Iustitia are provided with expression marks ('con gravità' and 'iratamente' respectively). The way in which Monferrato underscores the text is shown in Ex. 62.

Francesco Bagati wrote a dialogue ('Quae audacia, quae temeritas' (1658)) in which the Virgin (A) addresses Jesus (B) in favour of a repentant sinner (T). While the mediating role of Mary is a common theme in the repertory of the time, the implacable attitude of her son is rather exceptional. The language of Jesus resembles that of the allegory of Justice in Monferrato's piece: 'Iustitiam volo! Ferte mihi gladium Angeli flammeum! Fulmina prepara Michael!' So it takes a considerable time before the Virgin succeeds in softening her son's wrath. In the concluding terzet the characters step out of their roles, praising Mary and God (Jesus). Among the generally mediocre dialogues of Bagati the present one

Ex. 62

is by far the best. The composer's short-winded melodic writing is less in evidence here than in his other works. The final section, with its sustained cadential dissonances resolving only at the last moment, offers attractive music.

The same subject was treated by Giacomo Carissimi in his dialogue 'Doleo et poenitet me'. Yet in this case it is God (B) and Jesus (T) who hold strong differences of opinion about the fate of two sinners (SS). The question of how to reconcile their argument with the dogma of the Trinity is beyond the competence of a musicologist; it is better left to the theologians. Another problem, however, is of a purely musical order. Three viols playing in the middle and high registers are added to the vocal parts and this is quite unusual in the Italian repertory of motets and dialogues. Carissimi himself included one or two violins in a relatively small number of works; his use of the viol—extremely rare—is limited to the doubling or paraphrasing of the continuo. Since the only source of the work is a manuscript belonging to the Düben collection in Uppsala, it seems more than likely that the three viol parts were added by a German or Swedish scribe. A further argument can be adduced in favour of this supposition: nothing stands in the way of a purely vocal performance (with continuo). Nevertheless, my decision to include a transcription of the piece with viols in Part Two may be justified by the skilful treatment of the instruments. Moreover, the historical value of arranged works is in principle not less than that of original compositions. Both contribute to the image of the time.

Carissimi shows his mastery in several details, such as the litany-like monotony of the sinners' ritornello, which contrasts with the expressive singing of the tenor and bass (note the narrow range of the two sopranos in their ritornello: a minor third); further, the underlining of the words of Christ, 'but I have taken their sins and died for them on the altar of the Cross', by a striking turn from E major to G minor (bars 90ff.). The work is a worthy companion to the composer's better-known oratorio dialogues.[5]

Other Subjects

An undated work, preserved in manuscript, by Carlo Donato Cossoni, 'Ave Crux', does not fit into any of the previously discussed categories; the piece is neither a biblical nor a hagiographic nor even a moralizing dialogue. Instead, it takes a historical subject and as such it may be considered as a unique case in the repertory. The facts (or presumed facts) on which the composition is based are the following. In the year 614 the Persian king Chosroes II conquered Palestine, burnt the churches, massacred the priests, sold all the remaining Christians as slaves, and carried the Holy Cross with him. From 622 onwards the Byzantine emperor Heraclius fought various battles with the Persians who had invaded the Empire; in most of these he was successful. In 628 Chosroes was dethroned by his rebellious son Siroes and thrown into prison, where, in a literal sense, he had to face the slaughter of his other eighteen sons before he was cruelly dispatched himself. As for Siroes, after his accession to the throne he made peace with Heraclius, surrendering the territories which had been previously conquered, including Palestine. The Byzantine emperor himself brought the Holy Cross back to Jerusalem.

It is easy to see that the historical circumstances of the return of the Holy Cross were not spectacular enough to serve for an effective dramatic dialogue. Hence Cossoni (or his librettist?) invented a battle, the direct result of which was the recapture of the relic. The opening phrases are sung by the narrator, a personification of Santa Chiesa (S). However, right from the beginning it becomes clear that her role goes far beyond mere narration: 'Ave Crus, amabilis thesaurus, ave pacis ara! Quae bella te premunt, quis hostis invidus abscondit te? Quis contra tonantem audet pugnare?' (Hail thou Cross, lovely treasure, hail Altar of peace! By what wars art thou oppressed, what envious enemy hath taken thee away? Who will dare to fight the thunderer?) The means by which Cossoni underscores this emotional text include repetition of words, frequent changes of tempo, and word-painting ('pugnare' is depicted by trumpet-like motifs in the vocal part, while the continuo alludes to drums; see Ex. 2 above).

---

[5] A translation of the text of this dialogue is to be found in Dixon, *Carissimi*, 25.

From the concluding phrase of her little monologue it appears that, instead of *telling* us, the Santa Chiesa is *showing* us what is going on: 'Ecce terribiles bellatorum chori, ecce ordinatae militum falanges, ecce hostes!' (Behold the frightening host of warriors, behold the battle-array of soldiers, behold the enemies!) Then, after a general pause, the word passes to the Persian king (B), who boasts: 'En ego Cosdroas, immenso potens imperio dominator terrarum. Volo pugnare!' (Here I am, Chosroes, the mighty ruler with immense power over the world. I want to fight!) Heraclius (T), on the other hand, simply introduces himself as 'dux eterni Regis' (the general of the King of Eternity) before his voice is joined by that of Chosroes, who repeatedly shouts 'ad arma!' Subsequently the two adversaries sing either solo, in quick alternation, or together with short phrases in imitation.[6] While Chosroes continues to boast, reaching the highest note e' of the bass range with the outburst 'Exaltabo et ero Caesar altissimus', Heraclius, using descending motifs, tells him that he will be beaten.

Of course, the battle itself cannot be depicted in speech and, therefore, after an impressive general pause contrasting with the preceding tumult, the composer enlists the help of the Santa Chiesa to provide information about the outcome. She sings: 'The vengeance of Heaven, the wrathful God is near! Thou hast won, noble fighter, thou hast won, Emperor Heraclius!' Then the narrator steps out of character; she becomes almost a warrior herself, shouting 'Persequimini!' (Chase them!) Subsequently the attention shifts to the enemy and the principal key of F major changes into C minor. Chosroes moans over his defeat: 'Ah, miseri percussi sumus! Ecce fulminat contra nos Deus, ecce amittimus Crucem beatam, Crucem sanctam! Fugiamus in noctem damnatam.' (Ah, we are miserably struck! Behold how God fulminateth against us, behold how we lose the blessed Cross, the Holy Cross! Let us flee into damned darkness). The last word is again with the Santa Chiesa. She first speaks to God ('Thine enemies heard thee and are vexed') and then glorifies the victor.

During the whole dialogue there is a continuous alternation of slow and quick movement, sometimes at very close distance. This strongly enhances the dramatic intensity. The same is true of the short-winded phrases. While they would be out of place in a reflective motet, here they prove particularly effective.

To conjure up the atmosphere of a battle with the aid of only three voices and continuo is a *tour de force*, since the use of normal compositional procedures is restricted. It was probably for this reason that Cossoni considerably extended the customary concluding section in which the singers abandon their roles and together form a *choro d'angioli*. Here we find longer phrases, set homophonically as well as contrapuntally. A sevenfold cry 'Victoria!' is followed by a description of the enemy's flight, which the composer aptly depicts by means of a fugato based on a dactylic canzona-like theme. Subsequently there is

[6] See the musical examples in F. Noske, 'Sacred Music as Miniature Drama: Two Dialogues by Carlo Donato Cossoni (1623–1700)', *Festschrift Rudolf Bockholdt zum 60. Geburtstag* (Pfaffenhofen, 1990), 161–81.

a song of victory mentioning various musical instruments, such as *tympana*, *organa*, and *cymbala*, and after another praise of the Byzantine emperor the work concludes with an 'Alleluia'. This dialogue as well as its companion (*Il sagrificio d'Abramo*, to be discussed in Chapter 6) shows Cossoni to be a first-class musical dramatist.

Domenico Mazzocchi's *Dido furiens* and *Nisus et Euryalus* (both 1638), settings of fragments from Virgil's *Aeneid*, are exceptional in the Latin dialogue repertory because of their use of secular texts.[7] These pieces were written for the circle of *literati* centring around the Roman Aldobrandini and Barbarini families. Although the composer treated the epic poetry in a manner approaching the *stile rappresentativo*, there is no evidence for a staged performance. The presentation of a narrative text in the form of audible drama was not uncommon in Mazzocchi's time. Its most famous example is, of course, Monteverdi's *canto guerriero*, set to the description of the combat of Tancred and Clorinda from Tasso's epic *Gerusalemme liberata*. Yet there is a significant difference between the settings of Monteverdi and Mazzocchi. While the Venetian composer left the text unimpaired, as a result of which the principal role is that of the *testo*, the Roman gave preference to direct speech by omitting a considerable number of narrative verses. The latter procedure entailed occasional alterations of the original words for the sake of continuity. Yet in another respect Mazzocchi followed Monteverdi's example: his music observes the famous precept according to which 'harmony should be the servant of the words, not the mistress'. Hence the prevailing style is that of the recitative or semi-recitative. However, deference to the poetic language paradoxically resulted in the transformation of verses into prose. In the musical setting Virgil's enjambements are no longer recognizable as such and verbal elisions are ignored.

Dido furiens, treating the well-known episode in the fourth book of the epic (lines 296–512), includes only three characters: Dido (S), Aeneas (T), and Virgil acting as a narrator (B). The work is dominated by the solos of the heroine. Unlike seventeenth-century paraphrases of the subject, such as the libretto of Purcell's opera, the original text offers no instances of exchange of words at close distance. Nevertheless, by focusing upon Dido's state of mind and sacrificing for this purpose the text of her sister Anna as well as other 'superfluous' fragments, Mazzocchi succeeded in creating a truly dramatic scene. Expression marks ('concitato', 'arrabbiato') and abrupt changes of tempo (adagio, presto) eliminate the danger of monotony. Ex. 63 is characteristic. Compared with Dido's role, that of Aeneas seems rather pale, and this was undoubtedly the composer's intention. Virgil, on the other hand, is more than a mere narrator. Deeply involved in the dramatic situation, he uninhibitedly shows

---

[7] A third dialogued setting of a Virgilian fragment is in the composer's *Maphae I.S.R.E. Card. Barberini nunc Urbani VIII poemata* (Rome, 1638). This work was not available for the present study.

Ex. 63

his emotions in his melodic style. It is not by chance that melismatic passages are found in his part, rather than those of the principals. After Dido's death the work concludes with a short lament set for three voices.

The second dialogue treats an episode from the ninth book (lines 174–449). When, in the absence of their leader, the Trojans are besieged by the Rutulians, two young men, Nisus and Euryalus, set out during the night to cross the enemy's lines and seek Aeneas in Etruria. Unfortunately, they are intercepted. Nisus first escapes but then returns to save his friend. Finding him dying, he suffers the same fate.

Unlike the scene of Dido and Aeneas, this story contains much action; it also involves secondary roles. As there are only four voices available, the narrator Virgil has to share the bass with Aletes and Volscens, and the tenor with Aeneas's son Ascanius. However, the principals Nisus and Euryalus have each a part of their own. Since both of them sing soprano, their roles were undoubtedly meant to be interpreted by castratos. The work is in two parts. In the first, Nisus exposes his bold idea to his friend. Euryalus insists on accompanying him, and subsequently the young men ask permission for the expedition from the Trojan chieftains. Not only is this granted, but valuable gifts are also promised in case of their success. Euryalus, however, asks Ascanius to take care of his mother, calling this the most precious gift. The second part deals with the fatal outcome of the venture: their interception by the enemy, the killing of Euryalus, and Nisus's self-sacrifice. The musical procedures employed in this part are more varied than those in the scene of Dido. A 'chorus' of three (later four) voices, describing the agony of Euryalus (lines 435–7), is set to the Romanesca bass (Ex. 64). Subsequently, the narrator describes in emotional language how Nisus is killed by the Rutulian leader Volscens. The dialogue proper concludes with a eulogistic quartet set to the memorable lines 446–9:

> Fortunati ambo! Si quid mea carmina possunt,
> nulla dies umquam memori vos eximet aevo,
> dum domus Aeneae Capitoli inmobile saxum
> accolet imperiumque pater Romanus habebit.

In John Dryden's free translation:

> O happy Friends! for if my Verse can give
> Immortal Life, your Fame shall ever live:
> Fix'd as the Capitol's Foundation lies;
> And spread, where e're the *Roman* Eagle flies!

An optional addition is the *Lamento* of Euryalus's mother (481–97), set more or less in conformity with the principles of Nicola Vicentino (*L'antica musica ridotta alla moderna prattica* (1555)).[8] This particularly moving monologue,

---

[8] This 'appendix', which is headed by the precept 'Cantatur ut scribentur, rigorosè', was included by Arnold Schering in his *Geschichte der Musik in Beispielen*, no. 197, p. 241. The transcription is questionable, however.

Ex. 64

written in 'genere diatonico, cromatico ad enarmonico', includes a number of notes (E♯, B♯, and D♯) notated with a single cross (×) and meant to be sung a *diesis* higher than a normally sharpened note. The piece shows the almost complete disintegration of the modal system and may be considered the *nec plus ultra* of the 'seconda prattica'. Yet its compositional technique is anything but proto-tonal. Rather, one could speak of a blind alley. This, however, does not alter the fact that with his affective melodies and harmonies Mazzocchi brought out in splendid relief the poignant verses of the classical Roman poet.

5

The Latin Dialogue Outside Italy

The Italian repertory discussed in the previous chapters offers a fairly homogeneous historical picture. Starting mainly with the representation of biblical
scenes, it widens its choice of subjects after 1630 to include many hagiographic
and freely moralizing pieces. As for its musical characteristics, the semi-recitative,
dominant during the initial phase, increasingly becomes counterpoised to fragments written in aria style in the course of the century. In functional respects the
Latin dialogue kept its place within the liturgy of the Church or institutions
related to the Church.

For various reasons, the same cannot be said about the repertory of the *oltramontani*. In comparison with Italy, the number of Latin pieces composed in
Germany, Poland, England, the Low Countries, and France is rather small.
Moreover, the genre's function in relation to the Church is less clearly established, especially as regards Protestant regions in Europe. Neither the Lutheran
nor the Anglican Church rejected in principle the use of Latin words during the
service (as was the case with the Calvinist rite); this fact is attested by quite a
number of motets and anthems. Yet, unlike the motet, the dialogue, being written in *oratio recta*, needed more than a superficial understanding of the words.
This may be the main reason why, in an age witnessing a decline in the knowledge of Latin, composers preferred to set dialogued texts in the vernacular.
There are a limited number of English sacred dialogues, mostly preserved in
manuscript (Hilton, Ramsey, N. Lanier, Wilson, Purcell), but it was particularly in
Germany that the vernacular genre flourished, overshadowing that of the Latin
dialogue (Schütz, Scheidt, Schein, Ahle, Rosenmüller, Bernhard, and several others). As for the musical style, comparison of the 'foreign' Latin pieces with those
written in Italy meets with difficulties, not only because of the small number of
extant works but also because of the fact that most of them were composed after
1650. Yet a certain development is demonstrable in France; it leads from Du
Mont and his contemporaries to Charpentier.

A European region not considered in the pages that follow is the Iberian
peninsula. Although the subject is still largely *terra incognita*, we know that
Latin sacred music in Spain and Portugal was marked by an ultra-conservative
taste; the *stile concertato* was virtually rejected in this field. So it seems most
unlikely that any role dialogues were written in those countries.

Germany

Since in the seventeenth century there was no nation of that name, comparable to the kingdoms of Poland, England, and France, the term 'Germany' should be understood here as the German-speaking regions of Europe. Yet it seems that Latin dialogues were predominantly written in the northern part. I have not been able to trace any work of this kind in the Catholic Habsburg Empire. Although composers working in Vienna—among them the Emperor Leopold I— adopted Italian dramatic style in their sacred music, particularly in the *sepolcro*, they used exclusively Italian or (more rarely) German texts.

The earliest examples of Latin dialogues in Germany are five compositions by Daniel Bollius, who during the 1620s was organist and later became *Kapellmeister* to the Prince-Archbishop Johann Schweikard von Kronberg at Mainz. Four of these pieces bear the title *Dialogus*, and one is titled *Representatio harmonica*. Unfortunately, the manuscripts of all of them were destroyed during the Second World War; therefore they can only be discussed on the basis of the descriptions found in the writings of Winterfeld and Moser (the latter including a few musical examples).[1]

Bollius' dialogues treat subjects taken from the New Testament. The composer kept fairly strictly to the scriptural texts, occasionally adding freely invented words. This implies that narration predominates over direct speech. In some cases the narrator, called 'Evangelista' following the Lutheran custom, switches from one voice to another, a practice also encountered in Italy. But in addition, single characters are represented by more than one vocal part. In 'Intravit Jesus in quoddam castellum' (the story of the visit to Martha and Maria 'Magdalena' (Luke 10: 38–42))[2] Jesus's words 'Martha, Martha, sollicitata es erga plurima. Porro unum est necessarium . . .' are sung by the bass, but 'Maria optimam partem elegit' by the second soprano. Another example is found in the representation of the scene of the twelve-year-old Jesus in the Temple (Luke 2: 42ff.);[3] here the boy shares the soprano part with his mother. 'Filioli mei, diligite alter utrum',[4] on the other hand, is a dialogue, based on the Epistles of St John, in which the roles of the Apostle and two youngsters are consistently assigned to a bass and two sopranos respectively; therefore this piece seems comparable to Italian works of the time. The same is true of 'Domine, puer meus iacet in domo', representing the scene of Jesus (B) and the centurion's servant (T) as

[1] C. von Winterfeld, *Johannes Gabrieli und sein Zeitalter* (2 vols.; Berlin, 1834; repr. 1965), i. 205ff.; Moser, *Die mehrstimmige Vertonung*, 54ff.

[2] D. Bollius, 'Intravit Jesus in quoddam castellum', cited as MS 129c by E. Bohn, *Die musikalische Handschriften des 16. und 17. Jahrhunderts in der Stadtbibliothek zu Breslau* (Breslau, 1890).

[3] D. Bollius, 'Cum factus esset Jesus annorum duodecim', cited as MS 85, 1 by Bohn, *Die musikalische Handschriften*.

[4] D. Bollius, 'Filioli mei, diligite alter utrum', cited as MS 85, 3 by Bohn, *Die musikalische Handschriften*.

related in St Matthew's Gospel (8: 6 ff.).[5] From Moser's description of 'Ingressus Jesus perambulabat Jericho' (the story of the small Zacchaeus (Luke 19: 1 ff.)), it may be concluded that this piece is a dramatic motet rather than a true dialogue.[6]

Bollius dedicated his most impressive composition, the *Representatio harmonica conceptionis & nativitatis S. Joannis Baptistae*, to the archbishop.[7] This work, written before 1626, was 'composita modo pathetico sive recitativo in duos actus et sex scenas' and contains a great number of roles: Isaias Propheta, who sings the prologue set to the antiphon text 'Audite insulae' (A), Lucas Evangelista (T), Gabriel Archangelus (S), Zacharias Sacerdos (B), Elisabeth uxor Zachariae (S), Maria Virgo (S), Joannes Baptista (S), and four-part ensembles for the Populus, Vicini, and Cognati. An epilogue is sung by eight voices. It is understandable that modern authors have referred to this work as the first oratorio written in Germany. Yet such a description is mistaken. During the seventeenth century only those compositions that were intended to be performed in an oratory can properly be termed 'oratorio'. Rather, Bollius' *representatio* should be considered a Latin counterpart to the contemporary Lutheran *historia*, exemplified by the works of Schütz. As such it seems to be an isolated case.

Moser did not rate highly the musical value of Bollius' works, and the published fragments confirm this view. Historically, however, these early examples of Latin dialogues written in Germany are invaluable, not least because of the ample and varied use of instruments (cornetto, bassoon, theorbo, lute, strings, and organ—the latter's part occasionally being fully realized in Italian tablature).

Two other early pieces are by Christian Erbach jun. and Johann Erasmus Kindermann. Both of them were Lutherans. The younger Erbach probably studied with his famous father. It seems, however, that the latter's tuition was far from fruitful. The Canticle dialogue 'Fortis est ut mors' (1627), featuring a Sponsa (S) and a Sponsum (B), opens with a melodic phrase which, though in itself not unattractive, soon becomes tedious through endless repetitions on various degrees of the scale. The abuse of this device has already been mentioned in connection with early works by Cossoni and others. Erbach, however, goes far beyond these Italian examples. The words exchanged between the lovers, 'Ecce tu pulchrum/pulchra es, dilecte mi/dilecta mea', occur no less than eight times in the course of the piece. Yet the phrase is hardly recognizable as a vocal ritornello, since the same melody is used with slight variants for other words. Moreover, Erbach shows himself a poor harmonist; the parallel fifths and octaves in bars 82 and 135, not redeemed by contrary motion, are absolutely unaccept-

---

[5] D. Bollius, 'Domine, puer meus iacet in domo', cited as MS 88, 1 by Bohn, *Die musikalische Handschriften.*

[6] D. Bollius, 'Ingressus Jesus perambulabat Jericho', cited as MS 88 by Bohn, *Die musikalische Handschriften.*

[7] D. Bollius, *Representatio harmonica conceptionis & nativitatis S. Joannis Baptistae*, cited as MS 129 by Bohn, *Die musikalische Handschriften*.

able. We can only guess at Donfrid's reason to include this awkward piece in one of his collections.

Kindermann's dialogue about the Last Supper (1639; modern edition 1913) is quite another matter. Set to the verses 1 Cor. 11: 23–5 with additional words from Matt. 26: 26 ff., it features only the Evangelista (SS) and Jesus (B); strictly speaking, therefore, it is not a true dialogue. Yet, in its simple setting and adherence to the sober biblical text, the work represents the scene quite faithfully:

EVANG. Dominus Jesus Christus in ea nocte, qua traditus est, accepit panem et postquam gratias egisset, fregit deditque discipulis suis et dixit:

JESUS. Comedite! hoc est corpus meum, quod pro vobis datur. Hoc facite in mei commemorationem.

EVANG. Similiter et postquam coenavit, accepto calice, et cum gratias egisset, dedit illis et dixit:

JESUS. Bibite ex hoc omnes: hic calix novum testamentum est in meo sanguine, qui pro vobis effunditur in remissionem, peccatorum; hoc facite, quotiescumque biberitis, in mei commemorationem.

The opening phrase of the Evangelist is rendered with discreet, yet musically effective, devices of word-painting. The composer depicts 'traditus' (handed over, betrayed) by melismas in both voices, and 'fregit' (brake) by a hocket-like setting, the two syllables being separated by a crotchet rest. The words 'Dominus Jesus Christus' are set homophonically, but the remainder of the Evangelist's text is treated as canons at the lower fifth and fourth. The first solo of the bass forms a beautiful contrast with the foregoing. The words 'Take, eat: this is my body which is broken for ye; this do in remembrance of me' are set with repetitions of subphrases, each of which shows a heightened expression. Likewise, the following sections lack any routine treatment of the text. To take the word 'similiter' literally would seem trivial to a modern mind. In the Baroque era, however, such a procedure was anything but banal; hence the homophonic setting of the second part of the Evangelist's relation. The rendering of Jesus's words 'Drink from this, all of ye. This cup is the new testament in my blood, shed for ye for the remission of sins: this do ye, as oft as ye drink it, in remembrance of me' is more lively than that of his first statement. Yet the concluding melodic cadence is the same, a device which unifies the scene proper. This beautiful work concludes with a short terzet set to the text of the sacramental hymn 'O salutaris Hostia'.

Augustin Pfleger was among the pupils of Kindermann. His interest in dialogued music is attested by a cycle of seventy-two German 'cantatas', many of which are dramatic compositions with roles for biblical and allegorical characters. In addition, Pfleger wrote a number of Latin dialogues included in his first opus (1661). Although, unlike his teacher, he never journeyed south of the Alps, these pieces reflect the Italian musical taste of the mid-century. 'Jesu, amor dulcissime', a conversation between a soul (S) and Christ (B) about the renunciation of all earthly values as a condition for spiritual union, is partially set to verses. Hence the setting of the dialogue proper in triple-time aria style that approaches that of the Venetian bel canto. The concluding section, addressing the listener with the words 'Ergo, quicumque quaeritis vera gustare gaudia, terrena cuncta spernite', is set in common time; the use of the same musical material in both vocal parts symbolizes the union of the soul with Jesus.

Two biblical dialogues treat subjects which had long since disappeared from the Italian repertory: the fall of man and the Annunciation. The first piece, 'O pulcherrima mulier', does not deal with God's interrogation of Adam (as do the works of Corfini, Tomasi, and Cossandi discussed in Chapter 2), but with Eve's seduction by the serpent and the eating of the fruit from the forbidden tree. The composer fully exploits the dramatic possibilities contained in this scene. Changes of tempo, metre, and dynamics all contribute to a moving representation. The opening section, paraphrasing the conversation between the serpent (B) and Eve (S), is written in semi-recitative but ends in persuasive and joyful triple time: 'Comede, comede, o quam suave est' (Eat, eat, O how sweet it is). The soft repetition of the words 'suave est' proves particularly seductive, and so Eve answers with the same dance-like motifs: 'video, comedo!' Then the tempo changes to 'presto', the serpent enjoying his first success with coloratura-like exclamations: 'O quam hilaris felicitas, o quam felix jubilatio!' The ensuing seduction of Adam (A) by Eve is nothing but a shortened and slightly varied version of the opening section. In the seventeenth century the traditional depiction of Eve as 'a second serpent' was still readily accepted and Pfleger's treatment of this part of the story shows that he followed the custom. Once Adam has yielded to the temptation, the serpent indulges in an exuberant expression of his total triumph. The affinity of his melody with the previous motifs accompanying the fatal act is particularly striking (Ex. 65). From this point the serpent's joyful melodies, sung in fast tempo, alternate with adagio complaints by Adam and Eve about their future destiny. The work concludes with an ensemble for five voices describing the deceptively sweet taste of the apple: 'O pravum pomi gustum, o dulcedo nimis amara!'

Unlike this piece, which paraphrases Gen. 3: 1–6, Pfleger's Annunciation dialogue, 'Missus est Angelus', is an integral setting of the biblical text (Luke 1: 26–38). The work differs from the numerous compositions on the same subject, written by Italians during the period 1600–30, by its addition of two viola parts to the three roles: Evangelista (B), Gabriel Archangelus (S), and Maria Virgo (A). The treatment of these instruments is anything but spectacular. Yet, accompanying only the words of the narrator, they impart a particular colour to the scene. The archangel's role is by far the most important; his melodic language is expressive, especially in the description of Jesus's future greatness (vv. 32–3) and Elisabeth's pregnancy (v. 36). Mary's expression, on the other hand, is more reserved, but the soft repeat of the words 'quoniam virum non cognosco' (seeing I know not a man) forms a characteristic trait. Unlike most of his Italian con-

Ex. 65

temporaries, Pfleger used the opposition of forte and piano not as a conventional echo device but for specific purposes: 'persuasion' in the previously discussed piece, 'modesty' in the present one.

Practically the whole dialogue is written in semi-recitative style. Although this is also true of the Virgin's concluding words ('Ecce ancilla Domini, fiat mihi secundum verbum tuum'), her musical phrase is subsequently elaborated in an ensemble of all the voices and instruments. It is here that the two violas, departing from their secondary function, share the melodic material of the singers. A final touch, witnessing Pfleger's subtle procedures, is the double repeat of the word 'fiat' (let it be) in the last two bars.

We now arrive at the most important German composer of Latin dialogues, Kaspar Förster jun. His biography, recently completed and purged of mistakes by Barbara Przybyszewska-Jarmińska,[8] can be summarized as follows. Born in 1616 at Danzig (today Gdańsk) as the son of the homonymous bookseller and *Kapellmeister* of the Marienkirche, Förster studied with his father and (probably) with Marco Scacchi at Warsaw. It may have been the latter who advised him to complete his studies with Carissimi at the Collegium Germanicum in Rome. Förster's stay in this city from 1633 to 1636 was followed by a one-year sojourn in Florence; according to the papal nuncio in Poland, Mario Filonardi, he was there in the service of the Medici. In 1637 we find him engaged as a singer in the Warsaw court chapel. If we are to believe Mattheson's *Grundlage einer Ehren-Pforte*,[9] Förster began as a tenor but in the course of the years extended his range downwards, achieving the compass $a'–A'$, that is, exactly three octaves!

---

[8] B. Przybyszewska-Jarmińska, 'Kacper Förster Junior, Zarys biografii', *Muzyka*, 3 (1987), 3–19, with an appended thematic catalogue of Förster's forty-eight surviving compositions.
[9] J. Mattheson, *Grundlage einer Ehren-Pforte* (Hamburg, 1740), 21, 73 ff.

Even if we take this with a pinch of salt, the extremely low bass parts in his vocal works point to an exceptional voice. The stay in Warsaw, though interrupted by a second visit to Italy (1644–5), was the longest in the career of this otherwise restless composer (1637–51). By the end of this period Förster's fame had spread over large parts of Europe. In 1652 he was appointed to the prestigious post of royal *Kapellmeister* at the court of the Danish King Frederik III. Here he reorganized the chapel, engaging many singers from France, Italy, and Germany, as well as French violinists. Upon the death of his father in 1655 the Danzig authorities persuaded him to leave Copenhagen and accept the leadership of the chapel of the Marienkirche, offering him a high salary. Förster stayed two years in Danzig but then Italy tempted him again. Mattheson's account of his exploits as a *capitano* fighting for the Venetians against the Turks, a feat crowned with the award of the Order of St Mark, may seem fantastic. Yet the story should not be brushed aside, since a necrological poem, written some fifteen years later by the Danzig judge Ludwig Knaust, describes Förster not only as a great singer and *Kapellmeister* but in addition as a valiant man of arms. After this colourful intermezzo the composer resumed his duties in Copenhagen, staying there for a longer period (1661–7). In 1667 he returned via a detour to Danzig, passing through Hamburg (where he was received with open arms by Christoph Bernhard, Matthias Weckmann, and other musicians of that city) and Dresden (where he met Heinrich Schütz). Förster spent the last years of his life in Oliva near Danzig. He died in 1673 and was buried in the church of the Oliva monastery.

Förster's works match the cosmopolitan picture of his life. His vocal style is rooted in that of Italy, despite occasional French features. More international was his treatment of instruments, showing Italian and French as well as German traits. If we list him as a German composer (on the feeble basis that German was his native tongue), it must be said that he did not leave any composition set to German works. With the exception of a few Italian madrigals, all his surviving vocal works use liturgical or freely invented Latin texts. As for his religion, we find the same adaptable attitude. Though a Catholic, he seems to have felt no less at home in Protestant Copenhagen, Hamburg, and Dresden than in Warsaw, Rome, and Venice.

Förster left six biblical dialogues, all of which are preserved in the Düben collection in Uppsala. Their scoring ranges from three voices to six voices and ten instruments. The length of the works varies equally greatly. Two pieces run to *c*.150 bars and two others to *c*.500 bars. Stylistically, the six compositions show a remarkable versatility.

'Et cum ingressus esset Jesus navem' (ATB) is a dialogue about the pacification of the storm, mainly set to biblical words (Matt. 8: 23–7). As the number of roles exceeds that of the voices, the piece was probably written for an occasion with limited musical resources. The absence of instruments—rather exceptional in

Förster's sacred music—also points to this. The composer makes no attempt to arrive at a logical distribution of the parts. The narrator starts singing alto, but his description of the storm is brilliantly set for three voices in implicitly fast tempo (Ex. 66). The disciples' waking of the sleeping Jesus and the expression of their fear are likewise rendered by a solo passage in semi-recitative followed by a short terzet. The homophonic repeats of the word 'perimus' (we perish), separated by quaver rests, are particularly dramatic. Jesus's reproach ('Why are ye fearful, O ye of little faith') is assigned to the bass, and the ensuing description of his pacification of the storm to the soprano. A particularly original feature in the latter passage is the octave leap on the word 'magna' (Ex. 67). Marvelling at the performance of this miracle, the disciples again sing a 3. The concluding moralizing words about faith and confidence in God, set in triple time, are assigned to the tenor and subsequently repeated by the three-part ensemble. Although this work is the least characteristic of Förster's dialogues, it certainly shows his power as a composer.

If in the representation of the episode on the lake of Galilee the number of available voices was, strictly speaking, not sufficient, the opposite is true of 'Quid faciam misera', a dialogue about the woman of Canaan (Matt. 15: 22–8). The Gospel mentions only Jesus and the mother of the possessed girl. Förster, however, added to them a freely invented role for another female, who gives advice to the desperate mother and acts as her advocate. In addition, the score includes two parts for the viola da braccio, notated in the soprano clef. Since in the seventeenth century the term viola da braccio could designate any instrument of the four-stringed violin family, it remains unclear whether in this case the parts should be played by violins or violas; in my opinion, the latter are to be preferred. A complete transcription of the work is included in Part Two.

The dialogue starts with the lament of the mother and the advice of her companion to call on Jesus. This is written in expressive semi-recitative style. The accompanying instruments do not function as *concertato* parts; instead they 'realize' more or less the continuo. The mother's address to Christ is a beautiful example of semi-recitative evolving in an arioso. Jesus's first negative answer ('I am not sent but unto the lost sheep of the house of Israel') is again set in semi-recitative. After a second plea for help, answered by Jesus with the harsh words 'It is not meet to throw the children's bread to the dogs', the dialogue's text departs from that of the Gospel, reinforcing the dramatic tension:

Scriptural text	Dialogue text
And she said: Truth, Lord; yet the dogs eat from the crumbs which fall from their master's table	Am I a dog, am I a dog? Behold, I remain below thy table, eagerly longing for falling crumbs. And eventually they will fall.

These words are sung with particularly heightened expression (bars 93 ff.). Subsequently Jesus, now convinced of the mother's strong faith, heals the

Ex. 66

Ex. 67

daughter and the work concludes with a brilliant five-part setting of words borrowed from Psalm 116. Here the two instruments are treated as *concertato* partners.

Förster's dialogue about the parable of the rich man and the beggar Lazarus
('Vanitas vanitatum') keeps strictly to the biblical words (Eccles. 1: 2 and Luke
16: 19–31), albeit with a few abbreviations. The 'cast' consists of only three
roles: the Narrator (S), the Dives (T), and Abraham (B). The composer renders
the scriptural text in a rather dry, though occasionally expressive, recitative.
However, the work is considerably enriched by reflective terzets and instrumental ritornellos. Apart from two violins, the scoring includes three *viole in ripieno*
(actually two alto violas and one basso viola). The word *ripieno* does not refer
here to an 'orchestra' but simply indicates a non-concertizing function within an
ensemble.

The work opens with a brilliant *sinfonia*, followed by a three-part setting of
the famous verse from the Book of Ecclesiastes. The conversation proper is several times interrupted by a comment on the biblical text: 'Oh aeternitas!'
Nowhere in his dialogue does Förster approach the style of his teacher Carissimi
more closely than in this particular vocal ritornello. Another free text ('O mortales, quid fatales et inanes iuvant cura . . .') appears to conclude the composition. Yet this is followed by a literal restatement of the opening 'Vanitas
vanitatum'. Its final chord, lacking the third, conveys the emptiness of 'vanity'.

The most interesting aspect of the work is its layout. The dialogue proper,
including the narrative text portions, is set in an extremely sober way; it lays
stress on the intelligibility of the biblical words, behind which the composer
steps back. In the reflective sections and instrumental interludes, on the other
hand, he gives free play to his musical invention.

Opening words like 'Ah! peccatores graves somno delictorum, mortis exitio
conquassimus' hardly seem to point to a joyful Nativity dialogue. Yet they are
uttered by a shepherd in the field, acting as a spokesman for mankind in its sin-
ful state. In this way, the angels' message, 'Salvator vester natus est', occurring
some 140 bars later, is thrown into relief. The scoring is particularly rich and
colourful: SS (angels), ATT (shepherds), B (in the concluding ensemble), two
violins, two violas, spinet or viol, and basso viola. Moreover, the violins and violas are occasionally reinforced by four recorders, instruments that elsewhere

alternate with the strings. This is a practice which we will find again further on in this chapter in the scoring of a dialogue by Marc-Antoine Charpentier.

The opening *sinfonia* is in two sections, written in common time and triple time respectively. During the latter section the spinet twice has a short solo to be improvised on the continuo. After the pathetic lament of the shepherd (A), interrupted by desperate exclamations ('ah!'), Förster springs a surprise on us: a joyful instrumental interlude in the style of a French *noël*, played by the recorders (Ex. 68). Subsequently, the angels rouse the shepherds, who remain in their depression ('dolemus, ploramus . . .'). The fact that this is achieved by a quotation from the opening *sinfonia* shows the composer's endeavour to unify the work. An abbreviated repeat of the *noël* bridges the time needed for reaching Bethlehem. The dialogue concludes with an elaborate tutti-ensemble ('Gaudeamus omnes'), enclosing a terzet of the shepherds who address Jesus: 'Salve puerule, o dulcis Jesu, salve pax hominum, Salvator gentium, salve spes cordium salusque mentium!'

Ex. 68

The remaining two dialogues, both of which treat a subject from the Old Testament, are of considerable length. This is obviously the reason why in modern literature they have been referred to several times as oratorios. Since it is virtually certain that these works could not have been written for a service in the Roman oratory of S. Marcello, they should instead be considered extended dialogues, allowing *ripieno* performance of the vocal ensembles. In point of fact both compositions are described as 'motet' in the manuscript sources.

Giovanni Antonio Grossi's dialogue about the story of Judith, discussed in Chapter 2, proved rather unsatisfactory because of the missing part for the heroine's adversary. Unlike this work, Förster's 'Viri Israelite' (*Dialogus de Juditha et Holoferno*) represents Holofernes by a powerful bass (range $d'–C$). This contributes strongly to the dramatic qualities of his piece. The opening *sinfonia* shows rudimentary traits of the French overture: a slow introduction with dotted notes is followed by a lively 'fugal' section. The Historicus (A) does not restrict his role to mere narration; he addresses the Israelites in alternation with two 'speaking' violins (Ex. 69). The ensuing ensemble (SATB, violins, violas) expresses the fear of the besieged:

> O clades horrendae,
> Holofernes tremende!
> Squalida egestas,
> horrida tempestas
> premit incolas.
> Fremit civitas:
> Hosti serviamus
> ut invicem vivamus.
> Heu! cruda mors,
> predura sors.

Judith (S) strongly opposes this servile attitude:

> Quod est hoc verbum,
> ut introducatis superbum
> in civitatem sacratam
> Deo immortali dicatam?

Judith's journey to the besiegers' camp is dealt with summarily by the narrator, but her encounter with Holofernes is rendered in direct speech (recitative and semi-recitative). The Assyrian commander prematurely celebrates his victory with exaltation: 'Bibe, bibe et inebriare! Invenisti coram me gratiam beantem te'. The subsequent beheading of Holofernes is related concisely and a description of the heroine's return to Bethunia is dispensed with. What follows is a song of victory by the tenor, alluding to the enemy's consternation ('plangite dementes') and his flight. The role of the violins in this attractive section is particularly important. As in the Nativity dialogue, the composition concludes with an elaborate

Ex. 69

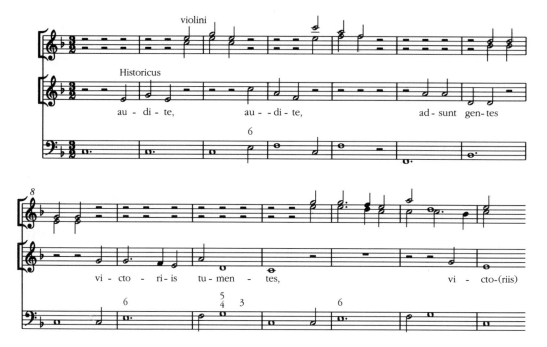

ensemble of all the voices and instruments; its text exalts Judith and praises God.

For all its attractiveness, this work is surpassed by 'Congregantes Philistei', a dialogue dealing with the combat of David and Goliath. The two subjects have much in common: a supposedly weak person (a woman or a boy), supported by strong faith and confidence, saves the people by gaining a victory over a much more powerful adversary. The circumstance favouring the representation of David's exploit above that of Judith is its enaction in a single place. This factor obviates the use of extended narration, inevitable in the story of Judith, and offers the opportunity of more direct speech, musically realized by various stylistic procedures.

The work, scored for four voices (SATB) and four stringed instruments (two violins, alto viola, and basso viola), opens with a short martial *sinfonia* that imitates trumpet motifs by means of the strings. The following solos by the Narrator (T), David (S), and Goliath (B), alternating with ensembles of the Philistines and the Israelites, offer more variety than the corresponding items in the Judith dialogue. In addition to sections in semi-recitative with motifs borrowed from the *sinfonia*, there are true arias such as Goliath's 'Nunc ubi pugnaces' (in da capo form, with two violins) and David's 'Tu gigas crudelis'. Even the narrator, glorifying David, sings in aria style. The combat itself, introduced by an abridged repeat of the *sinfonia*, is little more than a single stone's throw; it is described in a recitative of only six bars. Goliath's challenge and David's expression of confidence, on the other hand, occupy the greater part of the work. Of particular interest is the four-part ensemble of the Israelites expressing dismay and fear at the appearance of the giant; this 'chorus' is decidedly superior to the comparable item in the representation of the story of Judith. Since the text of 1 Kgs. (Samuel), Chapter 17, does not lend itself to being used verbatim, the work is set mainly to freely invented words with very few scriptural quotations. Some textual fragments depart greatly from biblical language, as, for instance, the following verses:

> Per Olympum, per quas stellas
> ergo tendit excelsa mens,
> super bellicas procellas
> victrix ibit humana gens.

This *Dialogus Davidis cum Philisteo* may be considered not only Förster's masterpiece but also one of the most accomplished works in the seventeenth-century dialogue repertory.

The question of where and when Förster's dialogues were composed cannot be answered with certainty. In view of the rich instrumentation Italy seems to be excluded, and it is also unlikely that the works were written during the composer's stay in Warsaw. The most acceptable place and time seem to be the second stay in Copenhagen during the years 1661–7. This conjecture is based on the maturity of the style and the availability of instruments at the Danish court. Also, the fact that the dialogues are preserved only in the Swedish Düben collection supports this assumption.

Finally a word must be said about Förster's main compositional quality: his versatile treatment of the chosen subjects. In the previous chapters of this book we have discussed many works of high musical value, including dialogues by Grandi, Donati, Mazzocchi, Graziani, Carissimi, and several other Italian composers. Yet, with the exception of Giovanni Antonio Grossi, none of these masters can match the astonishing variety of expression found in the works of this true cosmopolitan. His contribution to the genre is the proof that the *oltramontani* were anything but servile imitators of Italian models. More evidence of this will be encountered in other northern countries.

Poland

Like the Iberian peninsula in the south-west of Europe, the kingdom of Poland was a bastion of the Roman Catholic faith in the east. Yet, in regard to taste, the cultivation of Latin church music in the two regions showed different, indeed opposite, characteristics. Instead of the staunch conservatism displayed by Spanish and Portuguese composers, the Polish ones followed the modern trends of the time. Next to works written in the *stile antico*, they applied the *concertato* principle in both Masses and motets. Musical life in Poland centred around the royal chapel at Warsaw, among whose directors there were several Italians: Luca Marenzio (1595–8), Giulio Cesare Gabussi (1601–2), Asprilio Pacelli (1603–23), and Giovanni Francesco Anerio (1623–8). However, the most influential Italian master was the theoretician and composer Marco Scacchi, who directed the chapel from 1628 until 1649. Scacchi's doctrine, according to which each musical genre requires a distinct style with its own compositional technique, had a strong impact on Polish composers.[10] Yet the co-existence of stylistically independent genres was challenged by the anti-Italian Danzig organist Paul Siefert, who in his *Psalmen Davids nach französischer Melodey* (1640) had mixed various styles. The resulting polemical exchange of writings between Scacchi and Siefert is too well known to be discussed here. But it is conspicuous that Siefert also had a quarrel with the Italophile Kaspar Förster, sen., his own *Kapellmeister* at the Marienkirche. This points to the influence of Scacchi's ideas in German circles, a fact later corroborated by several treatises (Bernhard, Fux, Mattheson).

On account of the devastating wars over time, in particular the Second World War, only a small part of seventeenth-century Polish church music has survived. It may be assumed that, alongside Masses and motets, composers wrote sacred dialogues modelled on Italian examples. Yet only a single piece has been preserved: 'Audite mortales' by Bartolomiej Pękiel, the leading figure during the Polish middle Baroque period. Engaged as organist and later vice-director of the royal chapel, he eventually took over Scacchi's post sometime after the latter's return to Italy in 1649. As Kaspar Förster, jun., served as a singer in the chapel during the same years, the two men must have known each other very well.

'Audite mortales', an advent dialogue about the Last Judgement, scored for six voices and three viols, has been discussed by Polish musicologists from 1911 onwards.[11] Among the writings there is a penetrating analysis by Hieronim Feicht (1929), who in the same year also published an edition of the piece, based on the Berlin manuscript. Feicht and several others classified the work as an oratorio; this is once more an example of the uncritical adoption of the term, used anachronistically by Schering in his *Geschichte des Oratoriums*. It was only in 1968 that Pękiel's composition became identified as an example of the dia-

---

[10] M. Scacchi, *Cribrum musicum ad triticum Syferticum, seu Examinatio succinta psalmorum* (Venice, 1643). Other writings are mentioned in the entry 'Scacchi, Marco' by C. V. Palisca, *New Grove*, xvi. 542ff.

[11] Z. Jachimecki, *Wpływy włoskie w muzyce polskiej, Cz I: 1540–1640* (Italian Influences in Polish Music, Part I: 1540–1640) (Cracow, 1911); H. Feicht, '"Audite mortales" Bartłomeija Pękiela', *Kwartalnik Muzyczny*, 4 1929; id., 'Muzyka w okresie polskiego baroku' (Music in the Polish Baroque Period), in *Z dziejów polskiej kultury muzycznej* (From the History of Polish Culture) (Cracow, 1958).

logue genre in the excellent commentary by Zygmunt Szweykowski to his new critical edition, this time based on the more reliable Uppsala manuscript. The differences between the two sources are to be found mainly in the instrumental parts. Yet, although the viols are treated in the same manner as in Carissimi's 'Doleo et poenitet me' (see Chapter 4), in this case there is no reason to doubt their authenticity.

The text is partly a fantasy, partly a paraphrase of verses from the Book of Revelations. As has been pointed out by Szweykowski, its layout is almost symmetrical:

Sections	Textual content	Bars
I	Summons to the Judgement (Angel, A1°)	1–27
II–IV	Fear and terror of the sinners (T, S, AA)	28–83
V	Judgement of the sinners (Jesus, B)	84–107
VI	Lament of the damned (SSA2°)	108–38
VII	Judgement of the redeemed (Jesus, B)	139–65
VIII–X	Praise of God by the redeemed (A1°, T, SS)	166–223
XI	General praise (tutti)	224–51

If, adopting rhetorical terminology, we take section I as the *exordium* and section XI as the *conclusio*, then only section VI does not conform to the overall symmetry. Placed between the two Judgements (V and VII), it lies exactly in the centre of the work; yet as regards content it functions as the opposite of sections VIII–X (lament vs. praise).

The composer follows this scheme only in a formal sense. While the musical divisions do not overlap the textual sections—all of them close on perfect cadences—there is hardly any underscoring of the emotional contrasts. 'If we did not know the content of the text, we would not be able to determine whether a particular musical passage expresses fear, despair or joy.' Although basically true, this statement from Szweykowski's commentary to his edition seems too categorical. Occasional deviations from the 'neutrality' of the setting include the triple-time melody on Jesus's words 'Venite, properate, intrate in regnum caelorum' (bars 148–55) as well as the subsequent long melisma underlining the word 'exultate' (bars 158–65) and covering the whole range of the bass (d'–D). No one would associate these passages with fear or despair; they express unequivocal joy. Also remarkable is one instance of musical illustration. The angel's words 'Ecce tuba canit', likewise set to a melismatic melody (bars 15–18), are followed by a fanfare on the viols, which represent the biblical instrument, if not in tone colour, then by a tessitura fitting exactly that of an ensemble of three trombones. Szweykowski rightly points to the continuous change of scoring as a means to counter the general uniformity of expression: solo, duet, and terzet settings with or without viols alternate in the course of the piece. The overall impression is that of a most accomplished composition with reserved treatment of the text. In this respect it resembles more the Roman type of dialogue during the first three decades of the century than that of the north Italians.

In view of the many studies devoted to the work it would seem virtually impossible to add anything new to the known data, had I not come across two hitherto unnoticed German sources from the late seventeenth century. Although no longer extant, they throw light on Pękiel's reputation. In the introduction to his critical edition of Johann Krieger's church music Max Seiffert mentions that in 1860 this composer purchased the manuscript of 'Audite mortales', a motet a 9 by Giacomo Carissimi, from the Marienkirche in Halle.[12] Another source is found in the Ansbach inventory (1686).[13] Here the scoring of the work—again attributed to Carissimi—is more specific (6 voices and 3 instruments); moreover, mention is made of the key, F major. Although Andrew Jones included these two sources in his catalogue of Carissimi's motets,[14] there can be no doubt that in both cases the manuscript was actually a copy of Pękiel's dialogue. The mistaken attributions to Carissimi point to the esteem in which the work of the Polish composer was held in Germany, even though ten years after his death his name seems to have been forgotten.

England

Whereas in Poland Pękiel's single composition may be considered to represent a number of lost dialogues, in England the single extant piece, 'Heu me miserum' by George Jeffreys, seems to be a unique example.[15] As has been said in the beginning of this chapter, English composers virtually abstained from setting Latin dialogued texts, and even the vernacular sacred dialogue was anything but a favoured genre (among the relatively few works of this kind there is one famous piece, however: Purcell's 'In Guilty Night', treating the story of Saul and the witch of Endor).

In normal circumstances George Jeffreys would have exercised as much influence on the generation of Purcell as, say, Matthew Locke, or Pelham Humfrey, but for various reasons—political, perhaps also personal—he stayed for almost forty years outside the mainstream of English musical life. Hence, despite its importance, his work was known to only a few musicians of his time.

A visit to Italy may have preceded Jeffreys's appointment in 1642 or 1643 as

---

[12] J. P. Krieger, *21 ausgewählte Kirchenkompositionen*, ed. M. Seiffert (Denkmäler deutscher Tonkunst, 53; Leipzig, 1916), p. xvii.

[13] R. Schaal, *Die Musikhandschriften des Ansbacher Inventars von 1686* (Wilhelmshaven, 1966). The piece is mentioned on fo. 1015.

[14] Jones, *The Motets of Carissimi*, 14, 123.

[15] For this composer, see P. Aston, 'George Jeffreys and the English Baroque', diss. (University of York, 1970); id., 'George Jeffreys', *Musical Times*, 110 (1969), 772 ff.; id., 'Tradition and Experiment in the Devotional Music of George Jeffreys', *Proceedings of the Royal Musical Association*, 99 (1972–3), 105–15.

organist to Charles I, who during the Civil War kept his court in Oxford. Since he was a high churchman, there was no employment for the composer during the Commonwealth. Jeffreys spent these years (and the rest of his life) in the village of Weldon, Northamptonshire, under the protection of a Royalist country noble-man, Sir Christopher Hatton. At the time of the Restoration he probably felt too old to resume a professional career and so he remained an 'amateur' until his death (1685).

Even though the visit to Italy is undocumented, Jeffreys's familiarity with the *stile recitativo* and *concertato* may be taken for granted. Among his numerous manuscripts there are more than a hundred transcriptions of sacred works by Grandi, Carissimi, and lesser figures. It was his purpose to integrate Italian recitative-like declamation into the native tradition. This he fully realized in his late works, mainly Latin devotional songs, one of which is the dialogue.

'O me miseram', representing the scene of Mary Magdalene (S) and the angel (B) at the tomb of Christ, is preserved in no fewer than four manuscripts: three autographs and one non-autograph copy.[16] The text is a concoction of para-phrased verses from the gospels of Matthew (28: 5–10) and John (20: 11–18). The setting of the opening words, spanning a descending augmented fifth (D–B♭–F♯) and followed by a varied repeat with heightened expression (E♭–B♭–A), shows the Italianate handling of semi-recitative. Jeffreys even goes beyond his southern models by applying a procedure that may be called unique in the entire dialogue repertory. At the moment when she sees the angel, Mary Magdalene is literally dumbfounded and unable to pronounce the second sylla-ble of the word 'eum' (Ex. 70).[17] This feat of utter dramatic realism is achieved by a most radical application of the rhetorical figure of *abruptio*. The only other contemporary example known to me occurs in an *air de cour* by the Dutch diplomat Constantijn Huygens, like Jeffreys a skilled 'amateur' (Ex. 71).[18] Yet the two examples differ on an essential point. While Huygens employs the device to

Ex. 70

[16] I am indebted to Mr Jonathan Wainwright (Cambridge), who graciously provided me with his transcrip-tion of the piece and the bibliographical details in the List of Dialogues.

[17] As the truncated word appears in all the four scores, there can be no question of a clerical error.

[18] C. Huygens, 'Vous me l'aviez bien dit', *Pathodia sacra et profana* (Paris, 1647). Modern edition by F. Noske (Amsterdam and Kassel, 1957; rev. edn., Amsterdam, 1976).

Ex. 71

express intense emotion (the supervening rest functioning as an implied sob), Jeffreys renders the effect of a sudden change of situation (the rest depicting 'amazement').

The angel paraphrases the text of John, spoken by Jesus: 'Mulier, quo progrederis plorans?' His announcement of the Resurrection, on the other hand, is borrowed freely from Matthew: 'gaude et laetare: resurrexit Christus'. From here on the action is 'directed' by the angel. At his instigation Mary stoops down and looks into the sepulchre; yet, remaining unconvinced, she repeats her complaint 'Heu me miseram', which, now provided with chromatic notes, sounds all the more pathetic. Then the angel tells her to turn round. Seeing Jesus standing before her, she exclaims (in conformity with St John's Gospel): 'O Magister, o Domine!' The dialogue concludes with a joint expression of joy: 'Gaudeamus, exsultemus et laetemur, dicentes Alleluia!'

Throughout the work the music scrupulously serves the text with correct declamation. Yet the melodic writing is anything but dry; it shows beautiful contours. As the overall style, Italian rather than English, matches that of the best examples produced south of the Alps, we must regret that this dialogue remained a *unicum* among Jeffreys's sacred compositions.

The Netherlands

The Low Countries, covering the territories of the Spanish Netherlands and the Republic of the United Provinces, correspond more or less with today's kingdoms of Belgium and Holland. One would expect to find Latin dialogues mostly in the Catholic southern part and little or nothing in the Calvinist north, but the opposite is true. While there are more than a dozen pieces originating from the Republic, I have traced only a single composition written in the Spanish Netherlands (a dialogue by the Antwerp composer Anton Vermeeren). Since the sacred music of seventeenth-century 'Belgium' is still a largely unexplored field of research, it cannot be excluded that in the future more dialogues will be discovered in church archives. Yet it is unlikely that the genre was a favoured one in the Flemish as well as the Walloon regions. Strange as it may seem, the same can be said about the northern Netherlands. Apart from the Rotterdam organist Jan

Baptist Verrijt, who included a dialogue in a collection of motets, only one composer contributed to the repertory: Benedictus a Sancto Josepho, a Carmelite friar living in Boxmeer. His rich output, consisting mainly of Masses, litanies, and motets, includes no fewer than thirteen dialogues. Strictly speaking, Boxmeer was beyond the Republic's jurisdiction; the town, lying in the south-east of the country, belonged to a Catholic enclave, ruled by the counts of Bergh. Nevertheless, the composer must be considered as a Dutchman. Several of his works, published in Utrecht and Amsterdam, were owned and performed by municipal *collegia musica* in the Republic, for example that of Groningen.[19]

Vermeeren's 'Ah Domine Iesu, spes mea' (1661), mentioned above, features a woman (S) who, asking Jesus (B) how she should love him, receives the answer: '[With] your heart, your life, your soul'. The remainder of the text consists entirely of a mutual expression of love in the style of the Song of Songs, and so the dialogue lacks any element of tension. However, the composer added to the expertly treated vocal parts an instrumental trio (two violins and a basso viola), which plays throughout in *stile concertato*. It is the interaction of voices and instruments which compensates for the meagre textual content.

Jan Baptist Verrijt was born in Dutch Brabant and became nominally a Protestant when he was appointed organist and carillonneur in 's Hertogenbosch (1639). Five years later he moved to Rotterdam, having obtained a similar post at the main Calvinist church (St Lawrence's). Yet the Marian motets and Masses, which he published during his stay in the latter city, prove that he had remained *de facto* a Catholic.[20] The dialogue 'Fili, ego Salomon' is included in his only surviving collection *Flammae divinae*, dating from 1649. The way in which a single verse from the Book of Proverbs gave birth to a dramatic scene has already been described in Chapter 1. There can be little doubt that Verrijt borrowed the text from a piece by the Brescian composer Pietro Lappi (1614), which was reprinted in an Antwerp anthology (1622).

Lappi's setting is scored for six parts (SSATTB). The bass and the first soprano represent Solomon and his son, while the whole ensemble, interrupting three times the dialogue proper, addresses the congregation. It goes without saying that the *conclusio*, too, is set for six voices. The composer follows the textual layout scrupulously. Word repetition is avoided, even in the case of the son's emotional exclamations, such as 'Heu mi, peccavi!' The melodic and harmonic writing of the solo parts as well as the tutti ensembles is rather pedestrian and does not contribute much to the dramatic intensity of the scene. Taken as a whole, the quality of the composition is indifferent.

The three-part setting by Verrijt, a modern edition of which appeared in 1985, is in every respect superior to that of Lappi. The fact that the son's role is interpreted by more than one voice (AT) may be justified by the text of the conclud-

---

[19] See Spellers, 'Collegium musicum te Groningen'.
[20] For Verrijt's life and works, see Noske, *Music Bridging Divided Religions*, i. 111–32, 188–91.

ing section which addresses all young men: 'Deprecamur, o filii omnes, Christum, Dominum Salvatorum.' The altus and tenor sing either in homophony or in close imitation (including canonic passages). Repetition of words heightens the expression. The melodic language of Solomon is more sober, yet he summons his son with a particularly powerful melisma, spanning six bars. Comparison of the pieces by Lappi and Verrijt shows that the Italian simply *set* the text, whereas the Dutchman gave it a dramatic *interpretation*. An example is the son's exclamation, the only fragment in which Verrijt made a slight change to the original wording: 'Heu *Pater*, peccavi' instead of 'Heu *mi*, peccavi'. It is here that the son in a high state of agitation interrupts his father's repeated question: 'Peccasti?' Also, the syncopated setting of the vocal ritornello ('Intende mi . . .'), by means of which the work adopts a rondo-like form, is much more lively than that of Lappi, and the same is true of the expressive chromatic rendering of the concluding words: 'rogemus, *corde contrito*, ut nobis dimittantur'. Verrijt's dialogue is a model of its kind. The work may have been composed for a service in one of the 'hidden' Catholic churches in Rotterdam.

The composer 'Buns Geldriensis'—his baptismal name is unknown—was born in 1641 or 1642 in Gelder, a Dutch town outside the confines of the Republic. There he entered the Carmelite monastery in 1659, adopting the name of Benedictus a Sancto Josepho. At some time between 1666 and 1672 he was transferred to the monastery at Boxmeer, where he remained until his death (1716). His output consisted exclusively of sacred music: eight vocal collections and a set of church sonatas. The dialogues are included in Opp. 2, 3, 5, and 6, spanning a period of more than twelve years (from before 1672 to 1683).[21]

The most remarkable aspect of these works is the use of concertizing instruments. The 'normal' scoring is a trio of two violins with bassoon or a quintet of two violins, two violas (an alto and a tenor violino), likewise accompanied by a bassoon. In the following descriptions these two standard combinations will be mentioned as 'three' and 'five' instruments as appropriate; only deviant scorings will be specified. It was obviously for purely commercial reasons that in most of the dialogues the three lower instruments are indicated as optional. The fact that the violas often take part in the melodic interplay of voices and instruments makes them virtually indispensable. The same is true of the bassoon, which sometimes diverges from the basso continuo.

The *concertato* function of the instruments is already apparent in the two earliest works, included in the second edition of Op. 2 (1673; the first edition, probably dating from the late 1660s, is lost). Both pieces belong to the category of moralizing dialogues. 'O Jesu mi' (SSTB, 3 instruments) is a scene featuring two sinners plus one additional person and Jesus. The musical layout betrays a lack of experience in the handling of the dialogue genre. A large fragment, con-

---

[21] For the life and works of Benedictus a Sancto Josepho, see Noske, *Music Bridging Divided Religions*, i. 143–87, 191–7, 216–33, 257–62.

sisting of a phrase for Jesus ('Lavamini . . .') and the subsequent one for the sinners ('Tu potes, Domine . . .'), is repeated in its entirety after an intervening solo of the tenor. This weakens the dramatic impact. On the positive side, there is an impressive illustration of the text 'Ecce *stamus* ante te, considerantes et perpendentes vultum conscientiae nostrae'; these words are depicted by the unchanged harmony of B♭ major, spanning five bars. As for the instruments, they start with a *symphonia*—an item rarely missing in Benedictus' sacred music—and play short interludes between the phrases. In addition, they anticipate vocal incipits and accompany parts of the sung text, notably that of Jesus.

In 'Ego miser' the complaint of a sinner (A) is commented upon by two angels (SS), who subsequently give him the predictable advice to mend his ways and address Jesus. As a role for Jesus is lacking, the salvation can only be detected indirectly from the angels' text. This dialogue is dramatically more arresting than the previous one, not least because of the sinner's expression of amazement following his desperate utterances in the earlier part of the work: 'Quid est hoc? Me adhuc posse miserum salvari?' Its most remarkable aspect, however, is the replacement of the violas in the five-part instrumental scoring by two trombones. These instruments are not only used as middle voices in the harmony; they also play independently of the violins, forming a wind trio with the bassoon that accompanies the sinner's words: '[Dominus] qui me peccatorum miserrimum ad poenitentiam vocare dignatus est.'

The sole surviving copy of Op. 3 (1672) lacks the partbooks of the (first) soprano and the second violin. This affects in particular two of the four dialogues included in the volume. Since both of them are scored with soprano parts, a reliable reconstruction is out of the question. The cast of 'O quae laetitiae' includes two angels (SS) who invite the faithful to the Eucharist. Two sinners (TT), feeling themselves unworthy, do not dare to approach the altar; the angels, however, insist that they participate in the rite: 'Si timor abstrahit vos, amor allicit. Peccata solvite et cum fiducia comedite panem et bibite vinum.' The other piece, 'Ingemisco', likewise opposes a sinner (B) to an angel (S); its text describes the usual sequence of despair, conversion, and joyful salvation. The sinner's opening phrases exemplify the vocal style of Benedictus in his middle period. Word-painting (heaven, earth, hell) is embedded in expressive melodic contours (Ex. 72).

Two other dialogues, written without soprano parts, lack only the notation of the second violin, which can be rather easily reconstructed. The first of these pieces, 'Heu me miserum', represents a scene similar to that of 'O Jesu mi' from Op. 2: two angels encourage a desperate sinner. The fact that the work lacks the customary subtitle 'dialogus' is immaterial; its omission may be due to a lapse by the printer. The second piece, 'Mulier, quid ploras' (AB), is of quite a different character. After five examples of the moralizing dialogue, Benedictus at last tackles a biblical subject: the scene of Mary Magdalene at the tomb of Christ, based

Ex. 72

Heu mi - hi! et non est qui con - so - le - tur me. In - ge - mi - sco, mi - ser pec -

- ca - tor. Si o - cu - los in cae - lum fi - gam, mi - nan - - tem ju - di - cem

ti - me - o. Si ter - ram con-spi - ci - o, a - per - - - tum cer - no in - fer - ni

cha - os, ut me, ut me, ut me de-glu - ti - at. Nec cae - lum, nec ter - ram, nec

ma - re, nec a - er aut ul - la cre-a - tu - ra quae con-so-la-bi-tur mi - ser-ri-mam a - ni-mam me - am.

on John 20: 15–17. Conspicuously, both the scriptural words of Jesus and those of Mary are considerably extended with free additions.

	Gospel text	Dialogue text
JESUS.	Mulier, quid ploras?	Mulier, quid ploras, quid luges? Quid alto edis e pectore gemebunda suspiria? Ad quid toties tot iterati singultus? Quid ut perditio haec tot lachrymarum quibus terram rigas, caelum intonas irremedialiter? Pone moerorem ut blande arrideat facis ei tuae, laetabunda jam exurgens aurora.
MARIA M.	Domine, si tu sustulisti eum, dicito mihi ubi posuisti eum, et ego eum tollam.	*Ach, ach!* Quis dabit mihi moerorem hunc meum deponere, *quis dabit mihi pennas* quibus caelos conscendam, et inde referam medicamina moerori meo! Vel ipsos adjurem inferos, si invenerunt dilectum meum quem mihi abstulerunt, *ach*! nimium crudeles! O miseram me et infelicem! Domine, dicito mihi, quaeso, si tu sustulisti eum, et ego eum tollam.

The dedications of the composer's printed volumes, containing not only personal remarks but also examples of ingenious anagrams and puns, show him as an excellent Latinist. We may, therefore, safely assume that he himself wrote the free texts of his sacred compositions. In the present case the elaboration of the gospel text, including exclamations in Dutch and an allusion to Psalm 54: 7 (above in italics), may be considered too verbose. Yet on the other hand, this elaboration gradually heightens the tension, which reaches its climax in a rapid exchange of words (Ex. 73). The tension is maintained almost until the end of the dialogue. Now that Mary Magdalene has recognized Jesus, she wants to hold him, but he, speaking the gospel text, cannot allow this: 'Touch me not, for I am not yet ascended to my Father.' The work concludes with a short phrase sung by Mary: 'Behold, he whom I was longing for, I see him now.'

There is no reason to rate the musical qualities of this biblical dialogue higher than those of the previously discussed moralizing pieces. Yet the work is dramatically much more interesting. The same is true of the allegorical dialogue 'Posita in medio', contained in Op. 5 (1678; modern edition 1989). This composition is among the most remarkable in the entire dialogue repertory. Not only the great number of individual roles (four), matched by a five-part instrumental ensemble, but also the way in which they act may be considered exceptional. The 'cast' consists of a Soul (A), the Flesh (T), the World (S), and the Devil (B). After a short opening sonata (nine bars) the protagonist (the Soul) broods over her dilemma: 'Situated in the middle, I do not know which way to turn. Nature calls me to herself; grace recalls me to God. My imagination wants to follow nature, a just reasoning leads me to grace. But behold! There are the champions of nature, the chief enemies of grace, speeding to me. How shall I alone be able to resist them, without weapons?' Next, the Devil opens his attack: 'Why are you paralysed, why are you struck dumb? Where are you fleeing?' And, singing *dulce*, accompanied by two violins without violas and bassoon: 'Give yourself up, my dear, we are friends, do not flee.' The Soul soliloquizes: 'Woe! How can I resist, alone, unarmed?' Then follows a rather extended 'aside', a most rare occurrence in a sacred dialogue: 'Quick, Flesh/World/Devil, come quickly, we will offer her all kinds of desirable and delectable things.' The Devil takes the lead: 'You Flesh, sweet seducer, go first, and you World, father of lust, follow. Together reviving the desolate creature, you will deliver her to me.' Meanwhile the Soul repeats her words: 'How can I resist, alone, unarmed?' Both Flesh and World approach her with particularly sophisticated language. Flesh: 'God-loving creature, could you, fleshed, really live without me, without flesh in the flesh? Wherever you flee, you cannot avoid such a sweet consort. So stay forever with me in voluptuousness.' These words are matched by those of the World: 'God-loving, dulcet creature, how long, then, can you remain sensible in sensuality without senses? Enjoy charms, pursue riches, stay away from hardships, and you will be happy.'

Ex. 73

However, the Soul refuses to give herself up—she resists, calling for arms to chase away her seducers. In an animated ensemble with a quick exchange of single words (Soul: 'ad arma!'; Devil, Flesh, and World: 'pax!') the little drama approaches its denouement. Celestial bread serving as the sword of Gideon destroys Satan's camp. With the battle-cry 'Jo, jo! Vicit Leo de tribu Juda!' the Soul throws her adversaries into hell. The dialogue concludes with the expression of her triumph: 'and I, victorious, ascend to the stars'.

Throughout the work the instruments play an important part in the scene's dramatic representation, interrupting vocal phrases and subphrases, even verbal exclamations. In the above-described ensemble, for instance, they echo the Soul's cry 'arma', while the word 'pax', sung by her antagonists, remains unaccompanied. The mention of 'celestial bread on the altar' in the text points to the performance of the dialogue in connection with the Eucharist; this liturgical function is confirmed by its subtitle: 'De SSmo Sacramento vel de Tempore'.

The proficient handling of the *stile concertato* is also apparent in the remaining dialogues, all of which are contained in Op. 6 (1683).

Two moralizing pieces, both set for two sopranos and bass, show the great progress Benedictus had made during the past ten years. As in similar dialogues contained in Opp. 2 and 3, semi-recitative predominates. Yet the writing is much more varied and free from hackneyed formulae. The two works differ in several respects. 'O Salvator amantissime' (*Dialogus inter Christum et duas peccatrices*) is scored with five instruments. The vocal parts, which include quite a number of melismatic passages, are in perfect balance with those of the instruments. Although from a purely dramatic point of view the dialogue offers little interest, as a piece of music it is a brilliant achievement. The scoring of the other work, 'Quis dabit nobis' (*Dialogus inter Deum et duos peccatores*), with only three instruments, already points to a more intimate character. This moving piece, the style of which is expressive rather than virtuosic, shows the composer at his highest level. Another non-biblical subject is treated in 'Praecisa est' (SB, 3 instruments). Its title, *Dialogus inter beatum et damnatum*, is somewhat misleading. The two antithetical characters do not really converse. Instead, each of them speaks his mind in monologues, in turn as well as simultaneously. Only in a few instances do they seem aware of each other: for example, in the opening phrase ('Praecisa est velut a texente vita *nostra* et viam universae carnis *ingredimur*') and in the homophonically set fragment further on in the piece ('O sors, o sors *nostra*, quam diversa *nostra* sors!'). The most dramatic passage occurs towards the end, opposing exclamations of delight and despair (Ex. 74).

'Numquid ego' (*Ruina Luciferi*), a modern edition of which appeared some ten years ago, could be called a 'refused' dialogue. Like similar pieces written by Italian composers, it starts with Lucifer's display of megalomania (B). His antagonist—in this case God (S) instead of St Michael—asks him: 'Why, my angel, dost thou rise so high on the waves of thy pride?' Satan, however, ignores the

Ex. 74

question and continues to soliloquize, even after God's summons of the faithful angels to destroy the rebel and his companions:

GOD. Go to it, my angels, brave soldiers of the heavenly garden. Go and expel Lucifer.
LUCIFER. I shall ascend to heaven, and equal the Most High!
GOD. Be valorous and send the rebels flying.

The belligerent language is supported by martial instrumental motifs interrupting the vocal phrases. Subsequently Lucifer, recognizing his defeat, moans: 'I descend to the source of all misery. I who desired the supreme power over the heavenly kingdom; to gloomy terror and clamours fierce, to pallid shadows, to ardent fire, to weeping and fright in everlasting torment.'

The text of the concluding section, set for two voices representing the loyal angels, comments on the event and contains a warning, obviously addressed to the congregation: 'Such are punished; at Hell's gate their chains are prepared and the shackles of death. They miserably lament in their everlasting torment.' It is here that the restricted vocal scoring of the piece becomes unsatisfactory. One would expect a *conclusio* set for four or more voices, but, unlike Italian composers, such as Bonini and Donati, Benedictus never applied this procedure in his dialogues. However, the direction 'tutti' in the two partbooks suggests the addition of ripieno voices. Whether these were available in Boxmeer remains uncertain, but in any case the use of a chorus consisting exclusively of sopranos and basses makes a rather ineffective conclusion of the work.

Among the most accomplished dialogues in Op. 6 is 'O quam suave' (transcribed in Part Two). This is the scene of Christ visiting the sisters Martha and Mary 'Magdalene' in Bethany, recorded in St Luke's Gospel (10: 38–42). After the opening *symphonia* the redeemed Mary (S) and Jesus (B) express their mutual delight with many words (bars 1–97). This part of the scene, in which the three instruments have an ample share, shows a strongly developed handling of semi-recitative, which in this case almost adopts the style of the aria. Therefore the transition from common to triple time (bar 59), traditionally forming a contrast, is smoother than usual. Dramatic tension becomes apparent in the ensuing soliloquy of Martha (S); she worries about her domestic duties, which she cannot fulfil unaided. She then addresses Jesus: 'Lord, dost thou not care that my sister hath left me to serve alone? Bid her therefore that she help me.' This request is immediately countered by Mary 'Magdalene' with words not contained in the scriptural text: 'O Jesus, I am longing for thee, defend me.' The dispute is settled by Christ who, in accordance with the gospel, states: 'Martha, Martha, thou art careful and troubled by many things, but one thing is needful: and Mary hath chosen that good part, which shall not be taken away from her.' The remainder of the work consists of a terzet confirming these words and a likewise three-part 'Alleluia'. All this is written in attractive *concertato* counterpoint, involving both the group of three voices and that of three instruments.

Finally, there is a piece in which Benedictus departs from his standard instrumental scoring; in addition, it displays a different compositional technique in its opening section. This is 'Audite virgines', set for two sopranos and bass with two violins and two viols. Although the first viol can be replaced by a *violino terzo*, the notation of whose part is slightly different and therefore printed separately, the version with two viols must be considered the original one. As for the deviant compositional procedure, this is the traditional descending tetrachord used as a *basso ostinato*. It is above this pattern that Jesus (B) calls two virgins (SS) to a spiritual wedding. Ex. 75 taken from the opening section shows the application of this technique, one of the most remarkable achievements in Benedictus' *œuvre*. In particular it is the one-bar interruption of the triple-time *ostinato* by the first virgin's question in common time ('Quae vox?') which stamps this opening section as a highly original piece of music. The virtuosic interplay of voices and instruments continues in the following sections, the text of which alludes to the parable of the ten virgins ('accendamus lampadas nostras' (cf. Matt. 25: 1 ff.)). After an intervening sonata of nine bars, the dialogue proper ends with a solo for Jesus ('Ingredimini et mecum laetamini, quia dabo vobis gaudia aeternum duratura') and the expression of the virgins' delight ('O nos felices, nos beatas, quae transivimus per ignem et aquam et sic venimus in refrigerium!').

The subsequent part of the composition is set as a motet, the characters stepping out of their roles. From a dramatic point of view this produces an anticlimax, the more so as the concluding homophonic section is written in a rather routine-like way. This semi-dialogue is certainly not among the most balanced works in Benedictus' output. As regards its liturgical function, it may have been

Ex. 75

connected with the taking of vows by one or more nuns in the neighbouring Carmelite convent at Boxmeer (*Elzendaal*, founded in 1672), possibly on the feast of St Barbara (4 December), since this saint is mentioned in the text: 'Sancta Barbara, quae sponsum Jesum per vias ignotas, per carceres, per verbera, per ferrum, per mortem impavida secuta est'.

With his thirteen dialogues Benedictus a Sancto Josepho made an important contribution to the genre. Unlike the works of Förster and Charpentier, those of the Carmelite friar appeared in print. Their date of composition can be established approximately and this enables us to follow closely the development of his compositional style and technique. The most remarkable characteristic of the mature works is undoubtedly the integration of the instrumental parts into the vocal *concertato*. This procedure, which is only rarely encountered in dialogues written in Italy, strongly enhances the sprightliness and the expressive force of the musical conversation; it also lends a colourful background to the exchange of words. The fact that the musical style of Benedictus often strikes us as original may be explained by his relatively isolated position in the northern Netherlands. This left him no other choice but to find his own way.

France

A number of works by three composers will be discussed in connection with the Latin dialogue in France: Guillaume Bouzignac, Henry Du Mont, and Marc-Antoine Charpentier. The dialogues of a few minor figures (Thomas Gobert, Gabriel Expilly, Jean Granouillet de Sablières, Pierre Robert) are either lost or difficult to trace. It is unlikely, however, that they would add significant elements not contained in the works of the great masters. Although one recognizes a certain development in the course of the century, the historical picture is far from homogeneous, in particular during the early phase. The first examples of the insertion of dialogued 'scenes' into motets are found in the works of a composer who was virtually unknown in the country's capital. Moreover, despite the musical connections between France and Italy during the first half of the seventeenth century, the true role dialogue was introduced in Paris by an organist originating from the Low Countries.

The musical output of Guillaume Bouzignac (*fl.* 1620–40) is found in two large manuscripts, one preserved in Tours,[22] the other in Paris.[23] Only nine pieces in these sources bear the composer's name, yet it is generally assumed on solid grounds that more than one hundred anonymous works are also by him. Leaving apart four French chansons and a number of liturgical compositions (Masses and Vesper psalms), the greater part of his musical heritage consists of motets scored for five or six voices. Although none of these is entitled 'dialogus', many include conversational sections and could therefore be termed 'dramatic motets'.

Bouzignac's musical language is full of inner contradictions. Old-fashioned techniques and bold modern procedures are juxtaposed within individual works. Classical counterpoint in late Renaissance style alternates with *concertato* homophony and bits of recitative (all this without continuo). It often happens that characters are first represented by single voices, but subsequently the words of the former 'actors' are sung by an ensemble, a procedure incompatible with the concept of a role dialogue. The effectiveness of the conversational passages,

[22] Tours, Bibliothèque Municipale, Ms. 168. [23] Paris, Bibliothèque Nationale, Ms. Rés. V^{ma} 571.

mostly consisting of short questions and answers, is weakened by the ensuing 'choral' section, which, from a dramatic point of view, gives the impression of a *non-sequitur*. As a result of this, quite a number of works lack unity and are formally rather unbalanced. With his juxtaposition of heterogeneous styles Bouzignac certainly did not write in accordance with the rules established by his Polish–Italian contemporary Marco Scacchi. His great compositional qualities are found in the details rather than the overall structure. Bouzignac's care for the right expression of the text and the feelings of the individual characters as well as groups of persons is indeed most remarkable. A fragment from 'Unus ex vobis' (Ex. 76), a dramatic motet about the Last Supper (see Matt. 26: 21–2), represents the scene in a vivid and realistic way.

Ex. 76

Further examples of arresting dramatic rendering of the text are found in 'Ave Maria', 'Ecce aurora', 'Ecce homo', 'En flamma divinis amoris', and 'Noë, noë, pastores'. Although it is tempting to discuss these and other works in detail, this would exceed the scope of the present study. Moreover, Bouzignac's influence on his contemporaries was virtually nil. Working in various towns in the south and the middle of France, such as Narbonne, Grenoble, Carcassone (?), Angoulême, and Tours, the composer and his music were practically unknown in Paris. Even Mersenne, who mentioned Bouzignac with praise in his *Harmonie universelle*, knew him only by reputation.[24]

It was a musician of the subsequent generation who introduced the Italian *concertato* motet and the role dialogue in France: Henry Du Mont. Born in the neighbourhood of Liège in 1610, he studied at the choir school of Maastricht Cathedral from 1621 onwards. In 1630 he was appointed organist at the same church, but two years later he obtained an extensive leave to continue his study of composition with Léonard de Hodemont at Liège. After the death of his teacher (1636) he returned to Maastricht and it was only in 1638 that he emigrated to Paris. Here Du Mont worked as organist at the Church of St Paul and successively obtained distinguished posts in court circles. He was harpsichordist to the king's brother, the duke of Anjou, and later to the queen, Marie Thérèse. In 1663 he became *sous-maistre* of the royal chapel, a position which he kept until his retirement in 1683. He died in the following year.

Because of his distinguished career, Du Mont occupied an important place in the Parisian musical life. His sacred works (*petits motets*, 1652, 1657, 1668, 1671; *grands motets*, 1686) gave him the reputation of a pioneer of the genre. An additional contribution to his fame was the collection of neo-Gregorian chant, published in 1669 and today still included in the *Liber usualis*.

With his first opus, the *Cantica sacra* for 2–4 voices, Du Mont showed his familiarity with the Italian *concertato*, which he had undoubtedly obtained during the years of his study in the Low Countries. Not only the addition of a figured continuo part but also the affective melodic style and the harmonic devices are typical of the north Italian church music from the first half of the seventeenth century. It was not until 1668 that he ventured upon the composition of dialogues; the *Motets à deux voix*, published in that year, include six examples of the genre. Before we embark on a discussion of them, something must be said about the texts.

Apart from Domenico Mazzocchi's settings of fragments from the *Aeneid*, not a single dialogue written in Italy, Germany, Poland, England, or the Netherlands discloses the author of its text. In many cases we may assume that the words are by the composer himself, yet unequivocal proof is lacking. In France, on the

---

[24] M. Mersenne, *Harmonie universelle* (Paris, 1636–7, repr. 1963), bk. 7: 'Des instruments de percussion', prop. xxxi. Mersenne received the information about Bouzignac from his friend Gabriel de la Charloyne at Angoulême.

other hand, a few authors are known. I am referring in the first place to Pierre Perrin, a poet whose name is connected with the foundation of the 'Académie d'opéra' and the creation of the *tragédie lyrique*. His libretto of *Pomone*, the first work in this genre, set by Cambert, is mentioned in every textbook of music history. Less known are his collections of poetry, both in French and Latin, intended to be set to music. Not only the published *Cantica pro Capella Regis*,[25] but also a large manuscript offered to Colbert, 'Recueil de paroles de musique',[26] contain motet and dialogue texts. While Perrin's chansons strictly follow the rules of French versification, his Latin *cantica* are more freely conceived. Since the poet considered the motet 'une pièce variée de plusieurs Chants et Musique'[27]—these words obviously refer to the complex structure—he wrote rhymed verses with lines of different numbers of syllables. The following verse, entitled 'Dialogus Virgini', offers an example:

> O gloriosa Mariae viscera!
> O pretiosa Virginis ubera!
> Ubi Christus de coelo descendit,
> Ubi terrae salus pependit,
> Ubi creatus est mundi Creator,   5
> Ubi lactatus est mundi Salvator,
> Vos in aeternum laudamus
> Et canticis celebramus,
> O salutae et vitae fontes!
> O favi melle rorantes!   10
> O gloriosa Mariae viscera!
> O pretiosa Virginis ubera!

Two unidentified characters each sing a line in turn, with the exception of lines 7–8 and 11–12, which are sung by both. This text, set by Du Mont, is of course not a conversational dialogue but rather an exchange of statements. The same is true of another piece, 'Te timeo, Judex terribilis/Te diligo, Pater amabilis', in which two singers throw a light upon contrasting aspects of divinity. This 'dialogue' was also set by Du Mont. A third text by Perrin, 'Quare tristis es, anima mea?', is more dramatic. Starting with a quotation from Psalm 42: 5, it opposes an angel to a sinner. A setting by Thomas Gobert was included in a lost collection of motets and another by Du Mont in his 1668 volume.

Since Perrin wrote his *cantica* in verses, the remaining three prose dialogues set by Du Mont must be attributed to another author. According to Henri Quittard, who as early as 1906 published a valuable monograph on the composer, this might have been François Charpentier, a minor poet and member of

[25] Paris, 1665. The manuscript version, 'Cantiques ou paroles de motets . . .' (Paris, Bibliothèque Nationale), was written after 1660.

[26] Paris, Bibliothèque Nationale. Modern edition in L. E. Auld, *The Lyric of Pierre Perrin* (3 vols.; Henryville, Ottawa, and Binningen, 1986), pt. iii.

[27] Quoted from the preface of the *Cantica pro Capella Regis* (see n. 25 above).

the Académie française.[28] These pieces are 'Peccator, ubi es?' (*Dialogus angeli et peccatoris*), 'Dic mihi, o bone Jesu' (*Dialogus inter sponsum et sponsam*), and 'In lectulo meo' (*Sponsae sponso caelesti*).

Du Mont's two-part dialogues lack spectacular traits. The subjects, dealing with the sinner's repentance and salvation or the spiritual union of the soul with Jesus, are conventional. The settings follow the texts closely and the composer abstains from applying individual dramatic procedures. The dialogue proper is rather short and the homophonic or contrapuntal *conclusio* occupies the second half of the piece. The most remarkable quality of these compositions is the purity of the style, which manifests itself in the treatment of all the musical parameters: beautiful melodic contours, discreetly expressive harmony, impeccable counterpoint, participation of the continuo in the elaboration of melodic material, and limpid declamation of the words. Within the restricted dimensions of these dialogues 'Peccator, ubi es?' is the most lively and dramatic, not least because of the quickfire exchange of questions and answers by the angel and the sinner. It is perhaps not fortuitous that this piece was mentioned by Lecerf de Viéville long after the composer's death. Referring to Du Mont's motets, he wrote: 'On les achète encore, on goûte leurs graces naïves et ce dialogue d'un ange et d'un pécheur: *Peccator ubi es?* se chante encore avec plaisir . . .'[29]

James Anthony aptly characterizes Du Mont as 'essentially a miniaturist'.[30] Fortunately a most remarkable composition of large dimensions has been preserved in manuscript, the *Dialogus de anima* ('Oratorio (!) pro omni tempore') set for five voices and two violins. The first owner of this manuscript, Sébastien de Brossard, rated the work very highly: 'C'est une espèce *d'oratorio* très excellent ou dialogue entre Dieu, un pécheur et un ange, à la fin duquel il y a un très beau chœur *à 5 voc. cum 2 Violinis et Organo necessariis.*'[31] There can be hardly any doubt about the title 'Oratorio' being unauthentic; it was probably added by Brossard himself. Another error is the mention of a single angel: there are actually three angels. Moreover, the final 'chœur' should be read as an ensemble of five solo voices. Anthony's distinction between the ensemble of the angels as a *petit chœur* and the concluding section as a *grand chœur* is likewise unfounded.[32] Despite its larger scoring in comparison with the little dialogues contained in the 1668 volume, the work belongs to the category of the *petit motet*.

No one would contradict Brossard's praise of this dialogue: it can indeed be counted as a masterpiece. The textual content is conventional. A sinner (T) utters his complaint and his fear of eternal damnation, while God (B) shows

---

[28] H. Quittard, *Henry Du Mont: Un musicien français au XVIIe siècle* (Paris, 1906), 155 n. 1. As for 'Peccator, ubi es?', Charpentier's authorship is very unlikely, if not excluded, since this text was already set by Gasparo Casati in 1640 (*Il terzo libro de sacri concerti* (Venice)).

[29] Quoted by Quittard, *Henry Du Mont*, 161. [30] *New Grove*, v. 714.

[31] Remark in Brossard's *Catalogue des livres de musique*, Paris, Bibliothèque Nationale, Ms. Rés. Vm⁸ 21.

[32] *New Grove*, v. 713.

himself implacable. Urged by the angels (SAT), the sinner then humiliates himself and is subsequently redeemed. The concluding part contains the expression of joy by all the five voices. A remarkable feature of the work is its layout in three major sections, each of which is preceded by a *symphonia*. The first section consists of an exchange of phrases between the sinner and God. Typical of Du Mont is his discreet but effective use of chromaticism in the part of the sinner and the heightened emotional force of the repeated opening phrase with a slight variant: the flattened D in the descending melody on the word 'dolore' (Ex. 77).

Ex. 77

Apart from chromatic passages, the sinner makes ample use of descending diminished intervals (fourths and fifths). These are completely absent in the part of God, whose melodic language is diatonic. An outburst of divine wrath is expressed through a melisma of semiquavers ('quae devorabit impios'). The opening phrases of the two characters are rather long, but in the course of the section they become shorter, a procedure which enhances dramatic tension. The section concludes with a desperate outcry by the sinner: 'Quid faciam miser, ubi fugiam nisi ad te, Deus meus?'

After the second *symphonia* the vocal colour of the composition changes. Instead of the dark voices of the protagonist and antagonist we now hear the ensemble of angels forming a trio of soprano, alto, and tenor. First they divide the words among themselves:

> A. Deprecare in fletu . . .
> T. humiliare in pulvere . . .
> S. induere cilicio . . .

and then they sing together: 'et ne tardes converti ad Dominum, quia benignus est, et miserator. Et propitius fiet peccatis tuis.' This 'terzet' is set in a manner which combines homophony with *concertato* counterpoint. The sinner's subsequent conversion and plea for forgiveness are accepted by God, but, in order to maintain the dramatic tension for a while, the text applies a 'retarding' device that we have also encountered in the Italian repertory, for instance, in Fergusio's Magdalena dialogue (see Chapter 3). Unable to believe in his salvation, the sinner continues his complaint:

GOD. Et ego avertam faciem tuam a peccatis tuis, et omnium iniquitatum tuarum non recordabor amplius.

SINNER. Clamabo, clamabo die et nocte . . .

GOD. Et ego te exaudiam.

SINNER. . . . cum pungar in lacrimis . . .

GOD. Et ego te consolabor.

SINNER. . . . squalebo in ieunio . . .

GOD. Et ego te exaudiam.

SINNER. . . . Sedebo in pulvere . . .

GOD. Et ego te exaltabo.

It is only at this point that the sinner becomes convinced of his salvation.

During the first part of the concluding section the angels and God retain their identity, the sinner remaining silent (Angels: 'Gaudeamus coeli cives, gaudeamus angeli!'; God: 'Congaudete coeli cives, congaudete angeli!'). The remainder of the text, expressing joy in heaven at the repentance of a sinner, does not allow any more individual statements. The setting is for five voices sharing the same words, accompanied by two violins. Here Du Mont brilliantly displays his skill at applying the various procedures and techniques of the *concertato* style: homophony, imitative counterpoint, antiphonal treatment of groups of voices, etc. This splendid conclusion to the work marks him out as a great composer.

A word must be said about the character and function of the *symphoniae*. The first of these, opening the work, is solemn and impressive, but does not convey anything specific about the text of the first section. The second, on the other hand, is a moving piece full of chromatic melodic lines covering a fourth, descending as well as ascending. In this way the instruments depict the desperate mood of the sinner (Ex. 78). In the last bars, however, the function of this *symphonia* becomes forward-looking. Ascending diatonic scales in the violins and the organ continuo announce the entry of the angels (see the example in Quittard's book).[33] The third instrumental interlude expresses joyful sentiments in triple time and concludes with a 'petite reprise'. The symphonies in this work have a twofold function. Separating the sections, they serve to articulate the three-part layout; in addition, two of them comment on the text. Written in a Classical style, the purity of which foreshadows Corelli, they strongly contribute to the high quality of this dialogue.

---

[33] Quittard, *Henry Du Mont*, 172.

Ex. 78

Unlike his secular works, Du Mont's dialogues are written in an Italianate style. The semi-recitative predominates and is even recognizable when used for the two-part concluding sections. French features, such as non-illustrative ornaments in the melody and suspended ninths and sevenths in the harmony, are occasionally encountered, yet the general character of these works does not differ essentially from that of Italian contemporaries. The fusion of the two national styles, enriched with highly individual procedures, was the achievement of a younger composer: Marc-Antoine Charpentier.

Charpentier's career unfolded almost entirely outside court circles. Born in Paris in 1643, he studied for three years during the late 1660s with Carissimi at the Collegium Germanicum. After his return from Rome he was successively engaged by the devout and immensely rich Mademoiselle de Guise, the Jesuit church of Saint-Louis, and finally the Sainte-Chapelle du Palais. In all these three posts he had ample musical resources at his disposal. In addition, he was for a short time in the service of the Grand Dauphin and composed music for the theatre that later became known as the Comédie Française. Shortly after his death (1704) Charpentier's music fell into oblivion. The composer was rediscovered at the beginning of the twentieth century, but it is only since the 1950s that musicologists have started to unearth his large output (548 extant works). Thanks to several scholars and musicians—among whom H. Wiley Hitchcock (research) and William Christie (performance) deserve particular mention—we now know that Charpentier rather than Lully must be rated the greatest French composer of the seventeenth century. There can be no doubt that Charpentier surpasses his famous contemporary both in originality and versatility. A verse by Serré des

Rieux, published as late as 1734, recalls a few of the composer's individual procedures:

> Charpentier revêtu d'une sage richesse
> Des chromatiques sons fit sentir la finesse:
> Dans la belle Harmonie il s'ouvrit un chemin,
> Neuvièmes et tritons brillèrent sous sa main.[34]

The highly original treatment of dissonances is indeed a characteristic of the composer's musical idiom, notably that of the augmented triad which often resolves in another dissonant chord. In addition, mention should be made of bold simultaneous or successive cross-relations. But it is not only with dissonance that Charpentier demonstrates his individual treatment of harmony. The small phrase reproduced in Ex. 3 above, which is repeated several times on various degrees of the scale, starts on a 6/4 chord! As for vocal melody, the composer goes beyond Du Mont's care for correct declamation of the text. Particularly remarkable is the insertion of short rests in the musical phrases; they reinforce the cogency of the textual discourse.

Referring to music in general, Charpentier wrote in his little treatise *Règles de composition*: 'Le seule diversité en fait toute la perfection.'[35] As is testified by his entire output, he applied this rule at every level. Diversity of style and technique is found in genres, individual works, sections, and even within coherent phrases. Moreover, there is a great variety of vocal and instrumental scoring in cases where two or more works are set to the same text. All this does not lead to lack of balance and unity. The diversity enriching the composer's force of expression is controlled by his mastery of structure and sureness in handling forms. Italian and French stylistic elements are sometimes juxtaposed, yet they are mostly fused in such a personal manner that they are hardly recognizable as such. In point of fact Charpentier achieved 'les goûts réunis' long before this ideal became fashionable in France.

Charpentier's musical heritage is almost entirely preserved in autograph manuscripts which are today in twenty-eight volumes in the Bibliothèque Nationale, Paris. The greater part of his *œuvre* consists of Latin sacred music, which includes about two hundred motets, thirty-five of which contain roles for individual characters and groups of persons. In his *Catalogue raisonné* Hitchcock termed these compositions (H 391–425) 'dramatic motets', apparently for want of a better name.[36] Charpentier himself indicated a number of them as 'motet', 'historia', 'canticum', or 'dialogus' but many others lack such a generic title. In view of the variety of scoring and length, ranging from duets to eight solo voices, four-part chorus and orchestra, and from less than one hundred to

---

[34] Serré des Rieux, *Les Dons des enfants de Latone* (Paris, 1734); quoted from R. W. Lowe, *Marc-Antoine Charpentier et l'opéra de collège* (Paris, 1966), 15.

[35] Paris, Bibliothèque Nationale, Ms. nouv. acq. fr. 6355.

[36] Hitchcock, *Charpentier: Catalogue raisonné*, 291–316.

more than one thousand bars, it is indeed impossible to characterize these disparate works with a single term. I shall restrict myself here to the description of eight pieces belonging to the dialogue genre as defined in the first chapter. They are scored for 2–4 voices with and without instrumental parts. A composition of larger dimensions will be discussed in the next chapter.

'Famem meam quis replebit?' (*Dialogus inter esurientem, sitientem et Christum* (H 407)) starts in an unusual way. Normally, the questions of the hungry and the thirsty man (SS), answered by Christ (B), would be set in semi-recitative style. Charpentier, however, renders them in periodically conceived phrases. The questions occupying two bars each close on an imperfect cadence and the one-bar answers on a full cadence:

A major ——————▶ E major ——————▶ A major

ESURIENS. Famem meam quis replebit?	
CHRISTUS.	Ad me veni, fili mi.
SITIENS. Sitim meam quis replebit?	
CHRISTUS.	Ad me veni, fili mi.

The principle of alternation inherent in the textual subject (hunger–bread and thirst–wine) is also applied to the style, for example in a solo by Christ. The words 'Panem caelestem habeo', set with an aria-like melody in ternary metre, are followed by 'Si quis esuriet, saturabo eum', rendered in semi-recitative. Subsequently, this procedure is repeated, starting a third higher ('Fontem vitalem habeo . . . Si quis sitierit, inebriabo eum'). Further on in the piece one encounters a variant of this device. The doctrinal statement 'Caro mea est cibus et sanguis meus est potus', set as a true recitative sustained by semibreves in the continuo, strongly contrasts with the preceding dance-like melody in triple time, set to the words 'Venite, comedite et bibite'. This short sacramental dialogue, dating from the years 1682–3, concludes with a terzet in which the characters maintain their roles:

ESURIENS. O panis Angelorum, quam suavis . . .
SITIENS.   O vinum gratiarum, quam dulcis . . .
CHRISTUS. Dilecti convivae, comedite et bibite . . .

Unlike this work, 'Hei mihi infelix' (*Dialogus inter Magdalenam et Jesum* (H 423)) lacks inner contrasts. Yet it is a highly dramatic piece, paraphrasing the emotional encounter of Mary Magdalene (S) with the risen Jesus (A) as related in St John's Gospel (20: 11–17). The text is set in expressive Italianate semi-recitative style. Curiously enough the work concludes with a 'conflict' (Mary Magdalene: 'Liceat mihi, Domine, stigmata sacra tangere, osculari plagas tuas, amplecti pedes tuos'; Jesus: 'Noli me tangere'). Since the composition is only extant in a non-autograph manuscript, it might have been transmitted in an incomplete form. The fact that both vocal parts end homophonically on an unstressed beat corroborates this supposition.

Charpentier's dialogues include only a single piece treating the theme of the repentant sinner. This is 'Mementote peccatores' (*Dialogus inter Christum et peccatores* (H 425)). Apart from solo sections of Christ (Bar.) and each of the sinners (SS), the work contains a superb contrapuntal terzet sung by unidentified characters; its text ('Ah! cor durum, cor ingratum, cor saxeum . . .') seems to address not only the sinners on the 'stage' but also those in the congregation. Particularly impressive is the homophonic setting of the words 'Haeccine reddis Deo, haeccine reddis Creatori tuo?', interrupting the preceding contrapuntal elaboration of the melodic material. The ensuing complaints of the sinners are written in different styles. While the first of them expresses himself in semi-recitative, the second sings in the style of a French air, combining clear declamation of the words with a cantabile melody. The work concludes with a repeat of the terzet. In this case, too, the ending is somewhat unsatisfactory, since the sinners' salvation—almost obligatory in pieces of this kind—is missing. As with the Magdalene dialogue, there is no autograph of this work and so it could likewise be incomplete. On the other hand, mention must be made of a posthumous publication of 'Mementote' which appeared in a collection dated 1725 and gives us information about the author of the text: 'Les paroles sont du feu le reverend Père Commite, jésuite.'[37] Apart from a few slight variants, this printed version is identical with the manuscript copy. So the question must be left open.

A 'Prélude pour *Mementote peccatores*' (H 428a), scored for two unspecified treble instruments and organ continuo, anticipates the opening vocal melody. This is probably a later addition. Unlike the dialogue itself, this prelude is preserved in an autograph.

There are five Christmastide dialogues by Charpentier. Three of these use the same text, 'Frigida noctis umbra totum orbem tegebat'. The simplest of these pieces, H 421, dates from 1698–9. Scored for three high voices (two trebles and a normal soprano), it might have been written for performance in a convent. The soloists include the Historicus (S3), the angel (S1), and a shepherd (S2). In addition, there is a *chorus pastori*, an ensemble of three voices that is also used for text portions of the narrator. The dark opening section set in C minor strongly contrasts with the ensuing joyful air by the angel (C major). This is written in 6/4 time occasionally changing into 3/2, which produces a delightful effect. The chorus of the shepherds maintains the ingenuousness of this charming air: 'Surgamus, properemus, festinemus!' The passage of time needed for reaching Bethlehem is expressed by a long pause, indicated in the autograph: 'Faites en cet endroit un grand silence.' The work concludes with a *noël* consisting of three stanzas. The first half of each stanza is sung by one shepherd and the second half by all three of them, restating the melody in harmonization. The charm-

---

[38] *Meslanges de musique* (Paris, Ballard, 1725), 126–39, title: 'Dialogue en trio de Mr. Charpentier'.

ing simplicity of the tune fits the overall ingenuous character of the dialogue (Ex. 79). Two simultaneous notes (C–G) in the final bar of the lowest soprano part point to the use of vocal ripieni in this chanson.

Ex. 79

Throughout the piece the continuo part goes far beyond its function of harmonic *fondamento* of the voices. It serves for an improvised four-bar prelude to the opening phrase of the Historicus and plays short interludes during the discourse of the angel. In point of fact, the organ replaces the unspecified treble instruments in another setting of the text which Charpentier wrote more than twenty years before (H 393, dating from the mid-1670s). This version is more elaborate, in particular with regard to the ensemble of the shepherds which contains melismas on the words 'properemus' and 'festinemus'. There are also some slight textual variants. As for the concluding *noël*, the first half of the stanza is not sung by a solo voice but played by the instruments. Its melody differs from that of H 421 but is in the same vein. Typical of the active role of the instruments is their close connection with the melodic material of the voices. This involves even anticipation or echoing of passages set in recitative style (Ex. 80).

A third setting of the same text (H 414, 1683–5), scored for two solo voices, five-part chorus, and two treble instruments, contains an additional march in *rondeau* form, representing the journey to Bethlehem. For a description of this attractive composition the reader is referred to Hitchcock's excellent short monograph on the composer.[38]

The French style dominant in the Nativity dialogues is also applied in a work dealing with the visit of the Magi and written for the feast of the Epiphany: 'Cum

---

[38] H. W. Hitchcock, *Marc-Antoine Charpentier* (Oxford, 1990), 55–8. The description includes a table of the design, a graph of the piece's symmetrical structure, and a musical example (the tune of the 'noël').

Ex. 80

natus esset Jesus in Bethlehem' (H 395, dating from the mid-1670s). Its scoring
is for three vocal parts (SSB), representing the Magi, and two treble instruments.
Both Herodes (B) and the Narrator (S1 and S2) borrow the voices of the Wise
Men. In conformity with St Matthew's Gospel (2: 1–12) the partly paraphrased
text contains much narration. Yet the Magi appear individually in the role of the
Historicus, whose description of their gifts to the child is shown in Ex. 81. The
simplicity of the setting does not prevent Charpentier from applying bold harmonic procedures (Ex. 82).

Four hagiographic dialogues deal with the apocryphal story of St Cecilia
describing the conversion of her husband Valerianus and his brother Tiburtius.
Three of these with rich scoring (H 397, 413, and 415) also include the representation of the martyrdom of the Roman virgin and the two men. The text of the
fourth piece, 'Est secretum Valeriane' (H 394), confines itself to the conversions.
The setting is for three voices and two treble instruments. Cecilia's opening
phrase (high soprano) quotes the text of the Magnificat antiphon of the First

Ex. 81

Ex. 82

Vespers for the eve of the saint's day: 'Est secretum, Valeriane, quod tibi volo dicere: Angelum Dei habeo amatorem qui nimio zelo custodit corpus meum.' These words are moulded musically into a miniature ABA pattern. The first sentence is set as the antecedent of a periodical phrase, the consequent being played by the instruments (twice two bars). The second phrase ('Angelum Dei . . .') is also divided into two parts, but with a change of metre in the middle (3 + 5 bars). Subsequently, the opening melody is repeated literally. The use of this formal device, which combines periodicity with inner contrasts, is typical of Charpentier's sense of order within diversity. Other modes of expression follow. The conversation between Cecilia and her husband (B) is rendered in semi-recitative and the concise account of Valerianus' baptism by the narrator is set homophonically for three voices and two instruments. The conversion of Tiburtius (S) is dealt with more extensively. The enchanting perfume of the flowers given by the guardian angel convinces him; his christening, however, is symbolized by instruments only. The dialogue concludes with a terzet for the three characters, the text of which includes a glorification of celestial harmony. The words 'O concentus delectabilis' and 'O melodia suavis' clearly allude to St Cecilia, the patroness of music and musicians. Since this traditional association did not come into being until the beginning of the Renaissance, it is not mentioned in Jacopo da Varagine's relation of the story in his *Legenda aurea*.

As in several other works containing considerable narrative text portions, Charpentier termed the present composition 'canticum'. Obviously, he used the title 'dialogus' only for pieces in which a narrator either plays only a minor part or is altogether lacking. Such a work is 'homo fecit coenam magnam' (*Dialogus inter Christum et homines* (H 417)), possibly written during the early 1690s. A complete transcription of the piece is included in Part Two of this study. Its scoring is somewhat larger than that of the previously discussed compositions: four voices (A, T, Bar., B) and four specified instruments (two violins and two recorders). The narrator sings only the opening phrase, borrowing the voice of the second Homo (T). His words as well as those of Christ allude to the Last

Supper; yet the piece, set to a free text, is a sacramental rather than a biblical dialogue.

This composition demonstrates the composer's skill as a musical colourist. Throughout the work the violins and recorders are used in various ways, doubling each other or alternating in pairs at long as well as short distances. The first solo of Christ (Bar.) is accompanied by two recorders only, whereas the phrase of the third Homo (B), opening the concluding section, is set only with violins. The most striking contrasts of instrumental and vocal colour are found in the middle of the piece; here the violins play antiphonally between the separate statements of Christ ('Potest amor . . . audet amor . . . facit amor omnia') and are subsequently joined by the recorders between those of the men ('Credit fides . . . sperat fides . . . capit fides omnia'). The conversational part closes at bar 102; it is followed by an extensive *conclusio* containing delightful harmonic frictions. The work is certainly among the most attractive of the composer's dialogues.

Although in this chapter we have discussed only a small part of Charpentier's sacred dramatic music, the works described leave no doubt that in terms of quality they surpass those of his contemporaries and predecessors both inside and outside France. No other composer matches his originality, his finesse at handling melody, harmony, and tone colour, not to speak of his masterly application of structural procedures. Charpentier's outstanding contribution to the genre left no room for further development. As in Italy, the dialogue in France faded away after 1700.

6

The Sacrifice of Abraham: Six Dialogues Compared

The discussion of individual dialogues in the previous chapters was necessarily restricted to brief descriptions with a few analytical remarks. It is virtually impossible to draw the historical picture of a substantial repertory in any other way. In order to compensate, at least to some extent, for this disadvantage I shall use the present chapter to discuss six compositions in greater detail. Although differing greatly among themselves, these works have a common subject: the Sacrifice of Abraham as related in the first book of the Old Testament. This poignant story lends itself very well to the comparative analysis of different musical settings. Hardly any other biblical episode matches its dramatic impact. Moreover, its ethical implications, involving the delicate question of the incompatibility of natural human sentiments and divine will, offer a moral challenge to Christians as well as to Jews and Muslims.

The Biblical Source

The story appears in the twenty-second chapter of the Book of Genesis (1–18). Its text is the following:

Vulgate

1. Quae postquam gesta sunt, tentavit Deus Abraham et dixit ad eum: Abraham, Abraham. At ille respondit: Adsum.

2. Ait illi: Tolle filium tuum unigenitum, quem diligis, Isaac, et vade in terram visionis, atque ibi offeres eum in holocaustum super unum montium, quem monstravero tibi.

3. Igitur Abraham de nocte consurgens stravit asinum suum ducens secum duos iuvenes et Isaac filium suum; cumque

King James Bible

1. And it came to pass after these things, that God did tempt Abraham and said unto him, Abraham: and he said, Behold here I am.

2. And he said, Take now thy son, thine only son Isaac, whom thou lovest, and get thee into the land of Moriah; and offer him there for a burnt offering upon one of the mountains which I will tell thee of.

3. And Abraham rose up early in the morning, and saddled his ass, and took two of his young men with him, and Isaac

concidisset ligna in holocaustum, abiit ad locum, quem praeceperat ei Deus.

4. Die autem tertio, elevatis oculis, vidit locum procul,

5. dixitque ad pueros suos: Exspectate hic cum asino: ego et puer illuc usque properantes, postquam adoraverimus, revertemur ad vos.

6. Tulit quoque ligna holocausti et imposuit super Isaac filium suum; ipse vero portabat in manibus ignem et gladium. Cumque duo pergerent simul,

7. dixit Isaac patri suo: Pater mi. At ille respondit: Quid vis file? Ecce, inquit, ignis et ligna; ubi est victima holocausti?

8. Dixit autem Abraham: Deus providebit sibi victimam holocausti, fili mi. Pergebant ergo pariter:

9. et venerunt ad locum, quem ostenderat ei Deus, in quo aedificavit altare, et desuper ligna composuit. Cumque alligasset Isaac filium suum, posuit eum in altare super struem lignorum

10. extenditique manum et arripuit gladium, ut immolaret filium suum.

11. Et ecce angelus Domini de caelo clamavit dicens: Abraham, Abraham. Qui respondit: Adsum.

12. Dixitque ei: Non extendas manum tuam super puerum, neque facias illi quid- quam: nunc cognovi quod times Deum, et non pepercisti unigenito filio tuo propter me.

13. Levavit Abraham oculos suos viditque post tergum arietem inter vepres haerentem cornibus, quem adsumens obtulit holocaustum pro filio.

14. Appellavitque nomen loci illius 'Dominus videt'. Unde usque hodie dicitur: In monte Dominus videbit.

his son, and clave the wood for the burnt offering, and rose up, and went unto the place of which God had told him.

4. Then on the third day Abraham lifted up his eyes and saw the place afar off.

5. And Abraham said unto his young men, Abide ye here with the ass; and I and the lad will go yonder and worship, and come again to you.

6. And Abraham took the wood of the burnt offering, and laid it upon Isaac his son; and he took the fire in his hand, and a knife; and they went both of them together.

7. And Isaac spake unto Abraham his father, and said, My father: and he said, Here am I, my son. And he said, Behold the fire and the wood: but where is the lamb for the burnt offering?

8. And Abraham said, My son, God will provide himself a lamb for a burnt offering: so they went both of them together.

9. And they came to the place which God had told him of; and Abraham built an altar there, and laid the wood in order, and bound Isaac his son, and laid him on the altar upon the wood.

10. And Abraham stretched forth his hand, and took the knife to slay his son.

11. And the angel of the Lord called unto him out of heaven, and said, Abraham, Abraham: and he said, Here am I.

12. And he said, Lay not thine hand upon the lad, neither do thou any thing unto him: for now I know that thou fearest God, seeing thou hast not withheld thy son, thine only son from me.

13. And Abraham lifted up his eyes, and looked, and beheld behind him a ram caught in a thicket by his horns: and Abraham went and took the ram, and offered him for a burnt offering in the stead of his son.

14. And Abraham called the name of that place Jehovah-jireh: as it is said to this day, In the mount of the Lord it shall be seen.

15. Vocavit autem angelus Domini Abraham secundo de caelo dicens:

15. And the angel of the Lord called unto Abraham out of heaven the second time,

16. Per memetipsum iuravi, dicit Dominus: quia fecisti hanc rem et non pepercisti filio tuo unigenito propter me,

16. And said, By myself have I sworn, saith the Lord, for because thou hast done this thing, and hast not withheld thy son, thine only son:

17. benedicam tibi et multiplicabo semen tuum sicut stellas caeli et velut harenam, quae est in litore maris; possidebit semen tuum portas inimicorum suorum,

17. That in blessing I will bless thee, and in multiplying I will multiply thy seed as the stars of the heaven, and as the sand which is upon the sea shore; and thy seed shall possess the gate of his enemies;

18. et benedicentur in semine tuo omnes gentes terrae, quia oboedisti voci meae.

18. And in thy seed shall all the nations of the earth be blessed; because thou hast obeyed my voice.

Two aspects of this text are of particular importance in regard to its musical representation in dialogued form. The first is the mode of expression. Except for the summons (vv. 1–2), Abraham's instruction of the two young men (v. 5), Isaac's question and his father's answer (vv. 7–8), and the concluding words of the angel (vv. 12, 15–18), the whole sequence of events is communicated in narrative form. Because of this circumstance the author/composer is faced with the choice between two solutions: (*a*) the addition of a role for a *historicus*; (*b*) the replacement of indispensable narration by direct speech and the omission of the rest of the narrative text. Both solutions have inherent disadvantages. A historicus subverts the spontaneity of expression, in particular at the crucial moments in this miniature drama (see, for instance, v. 11: And the angel of the Lord called unto him out of heaven, and said . . .). Omission of large portions of text, on the other hand, may lead to an incoherent rendering of the story.

The second aspect is the content. If we understand the concept of 'drama' as the intergeneration of action and emotion, then it must be said that in this case the biblical text provides a clear picture of the action but makes no mention whatever of the emotional experience of either Abraham or Isaac. In other words, while the events are explicitly related, the personal sentiments of father and son are left to the reader's imagination. Music is, of course, a particularly effective means of communicating implicit emotions, yet the dry scriptural words act as a restraining factor in the realization of this aim.

The discussion of the six pieces will show how their composers dealt with these problems. Mention must be made, however, of practical circumstances limiting their individual freedom: for instance, the fact that in some cases they had only the required minimum of three voices at their disposal; this in itself excluded the addition of a narrative role. Another restriction was the interdiction encountered by certain composers forbidding the deviation from the biblical words, only omissions being allowed. Finally, it is worth suggesting a possible motivation for Lenten performance of this dialogue: the appearance of Abraham in the Epistle for the fourth Sunday in Lent (Gal. 4: 22–31) means that his example must have been a frequent area of Lenten meditation, perhaps during the following week.[1]

Composers, Musical Sources, and Scoring

Together, the six works (three of which have appeared in a modern edition) cover the greater part of the seventeenth century (1615 to the 1680s), and the composers roughly represent the diffusion of the dialogue inside and outside Italy. Three of them (Giovanni Francesco Capello, Giovanni Antonio Grossi, Carlo Donato Cossoni) worked in the northern part of the peninsula, two (Abundio Antonelli, Giacomo Carissimi) in Rome, and one (Marc-Antoine Charpentier) in Paris. Four pieces are preserved in manuscript (Antonelli, Carissimi, Charpentier, Cossoni), one in print (Capello), and one in both manuscript and print (Grossi). Carissimi's composition is an oratorio dialogue, almost certainly written for the confraternity of the SS. Crocifisso and intended for performance on a Friday in Lent; the same may be true of Antonelli's piece. The remaining works are church dialogues. Only one of these (by Cossoni) provides an indication of its function within the liturgy: 'Per il Santissimo [Sacramento]'.

Unlike the Capello dialogues discussed in Chapters 2 and 3, which come from his Op. 5, 'Abraham, Abraham!' is included in Op. 7, *Motetti e dialoghi*, a collection published in Venice in 1615. The scoring of the work is for three voices: S (Isaac), T (Angel), and B (Abraham); in addition there are four parts for unspecified strings. The piece is probably the earliest dialogued representation of the story. A modern edition appeared in 1931.

Antonelli's relationship with the Roman confraternity of the Crocifisso is only conjectural, since its archives are incompletely preserved. Moreover, the manuscript of his Abraham dialogue, extant in a nineteenth-century copy by Fortunato Santini, lacks a date (the work was probably written during the 1620s). Yet, since the story of Abraham's sacrifice was a typical Lenten subject, well suited to oratorical performance, it is likely that the piece was composed for a gathering in the S. Marcello oratory. The scoring includes the voices of God (B), Abraham (T), Isaac (A), and the Angel (S), all of which are also used for the concluding double-choir ensemble (a 8). In addition, three instrumental parts are written for violin, lute, and theorbo.

Two other works date from the middle of the century. The autograph of Grossi's 'Abraham, Abraham!' is included in one of his collections of sacred compositions, dated 1660–2 and preserved in the Cathedral Archives in Milan. The

---

[1] See Dixon, 'Oratorio o motetto?', 203–22.

printed version, almost identical with the manuscript, appeared in the composer's Op. 7, *Il terzo libro de concerti ecclesiastici* (1670). Although the last-named publication contains several motets with *sinfonie*, the present dialogue is scored for voices only: S (Angel), S (Isaac), and B (Abraham). Its text deviates greatly from that of the Bible.

Carissimi's *Historia di Abraham et Isaac* exists only in manuscript copies that give no indication of the date of composition. On stylistic grounds it may be assumed that the work was written during the 1650s or early 1660s. Unlike Grossi, Carissimi represented the story largely in accordance with the scriptural text, adding to the dramatis personae, Deus (B), Abraham (T), Isaac (S), and the Angel (A), an important role for the Historicus (T2). However, most of the conversation between father and son is freely invented. All the five voices join in a concluding quintet (chorus). Together with *Job* and *Jephte*, this work belongs to those oratorical compositions by Carissimi lacking concertizing instruments. A modern edition is included in the *Complete Works*.

Il sagrificio d'Abramo by Cossoni is also undated. In view of the mature style, lacking the hackneyed formulae encountered in the composer's printed dialogues (1665, 1670), I would guess that the piece was written during the years when he worked as *maestro di cappella* at Milan Cathedral (1684–93). The autograph of the composition was bequeathed with many other works to the Collegio of Bellinzona, administered by the Benedictines; today the manuscript is in the library of the monastery at Einsiedeln. Like Grossi's setting, Cossoni's piece is scored for three voices only, S (Angel), A (Isaac), B (Abraham), and contains many non-biblical text portions. A modern edition of this dialogue has been included in a recently published article by the present author.[2]

Sacrificium Abrahae by Charpentier (H 402 and 402a) is not only the longest of all the six dialogues (461 bars); its scoring is also the most opulent: SSAATTBB soli and four-part choir, occasionally doubled by instruments that include two violins; these also play a ritornello twice and have individual concertizing parts in the concluding ensemble. One reason for the length of the work is the inclusion of the previous story, as told in Gen. 21: 1–8. Therefore, the 'cast' includes Sara (S), the Angelus (S), Isaac (A), Abraham (T), Deus (B), and the Historicus, whose words are sung by various voices and ensembles. According to Hitchcock, the composition dates from the years 1681–2.[3]

Antecedents and the Summons (Gen. 21: 1–8 and 22: 1–2)

As has been mentioned above, Charpentier prefaces the story proper by a conversational and narrative representation of the first eight verses of Gen. 21,

---

[2] Noske, 'Sacred Music as Miniature Drama', 161–81.
[3] Hitchcock, *Charpentier: Catalogue raisonné*, 299–300.

describing Isaac's birth and childhood. Since among the six composers he was the only one to do so, this introductory part will be dealt with summarily. The text, though partly paraphrased, does not deviate from the scriptural reading; the order of the verses is changed, however. The opening recitative of the Historicus, 'Cum centum esset annorum . . .', combines vv. 1, 2, and 5 (the birth and Abraham's age). Subsequently, Sarah recalls her previous scepticism at the divine announcement of her motherhood (vv. 6–7; cf. Gen. 18: 10ff., and Grossi's dialogue 'Heu! Domine, respice et vide' (discussed in Chapter 2 and transcribed in Part Two)). Sarah's solo is set partly as an 'aria' spanning only six bars, partly as a recitative. The ensuing attractive duet of the parents, 'Gaudeamus igitur', uses freely invented words. Its first part (A), written in double counterpoint, is followed by a short intermediate section (B) in which the voices of Abraham and Sarah alternate. The duet concludes with a varied restatement of A. Next a chorus, representing the Historicus, sings the verses 4, 3, and 8 (in that order); these describe successively the circumcision, the name-giving, and the feast on the day when Isaac was weaned. All this is set in strict homophony, serving the intelligibility of the text. Towards the end the added words 'et cum Isaac factus esset iuvenis, tentavit Deus Abraham' announce the summons in the first verse of the twenty-second chapter: 'Abraham, Abraham!' These are also the opening words in the dialogues of Capello, Antonelli, Grossi, and Cossoni; alone in Carissimi's composition are they preceded by a slightly altered version of the biblical narrative, sung by the Historicus. Introductory *symphoniae* occur in the works of Antonelli and Capello; the latter's 'piece' will be repeated later with semantic implications. While Capello's strings play homophonically, Antonelli uses his heterogeneous ensemble of violin, lute, theorbo, and 'organo concertato' for contrapuntal elaboration of melodic material.

Strictly speaking, only Capello's setting of the summons is in accordance with the scriptural text; all the others contain additions, omissions, or variants. Both Antonelli and Carissimi dispense with Abraham's response 'Adsum'. Cossoni's text, on the other hand, represents the scene with great imaginative power:

ANGEL.	Abraham, Abraham! [*bis*]	Abraham, Abraham! [*bis*]
ABR.	Quae vox de caelo sonat?	What voice sounds from heaven?
ANGEL.	Angelus Dei vocat.	The angel of God calls thee.
ABR.	Adsum. Quid fieri iubes?	Here am I. What dost thou want me to do?

One cannot but admire the realistic setting of these words. Obviously, Cossoni started out from the idea that, even for a biblical patriarch, a voice addressing him from heaven was anything but a trivial event. Hence his implied dumbness, rendered by a rest of two bars during which the continuo plays alone. The angel calls Abraham a second time twice by his name, using the same signal-like motif (a broken triad). Abraham's first question, including a long diatonically descending melisma on the words 'de caelo', is answered by the angel, who likewise emphasizes the word 'Dei' by a melisma. Then the tempo changes from 'largo' into 'svelto' (synonymous with 'allegro'); Abraham, emerging from his stupor, repeatedly sings the word 'adsum', using a lively motif echoed by the continuo, and puts himself at the disposal of the divine wish. Not only Cossoni but also Capello and Grossi replace God by an angel; the last-named composer cautiously justifies this substitution by the words 'Angelo (voce di Dio)'. As Cossoni had done, Carissimi, Grossi, and Charpentier set the call 'Abraham, Abraham!' to a triadic motif. Capello uses instead a diatonically descending melody, both for the call and the answer. Additional words occur in Grossi's setting: 'Adsum Domine. Effare, impera quid vis' (Here am I, Lord. Speak, order what thou wanteth to be done).

The instruction (v. 2) is set by all the six composers in a 'spoken' way, ranging from pure recitative (Carissimi) to a style approaching that of an arioso (Cossoni). Capello quotes the descending melody, previously set to 'adsum', in the continuo. This may be a purely musical procedure, yet it could also be interpreted as a subtle means of depicting Abraham's close attention while listening to the divine command. Grossi's text, unlike that of the other composers, deviates widely from the biblical reading, reducing it to the following words: 'Tolle puerum, quem amas, Isaac, in montis apicem, ut offeras mihi in holocaustum.' His setting, too, is quite remarkable; it contains a vehement melisma and includes the rhythmic design which I have called the 'Piacenza formula' (Ex. 83).

Ex. 83

Abraham's reaction to the words from heaven is, not surprisingly, expressed by Cossoni. The patriarch soliloquizes: 'Do not hesitate, my heart. The command is terrible, but it is the will of God.' These words are set with continuous alternation of slow and quick movement, an effective means of depicting strong emotion. Grossi's text also comments on the divine order. Here Abraham speaks to the angel: 'I am ready to serve thee, to adore thee. No paternal love shall contradict thee, nor shall my son's anguish thwart me.' Curiously enough, none of the other dialogues makes mention of Abraham's sentiments. One would expect Charpentier, in particular, to avail himself of this opportunity to depict the father's state of mind.

The Journey (Gen. 22: 3–8)

The treatment of this part of the story differs greatly among the six composers. Abraham's preparations for the journey (v. 3) remain unmentioned in the works of Capello and Antonelli. Carissimi's Historicus describes them succinctly in his own words, omitting the fact that father and son will be accompanied by two young men. Charpentier, on the other hand, renders the scriptural text almost integrally, his narrator being represented by an ensemble of three solo voices (ATB). Neither Grossi nor Cossoni is concerned with the material preparations; instead, they draw attention to a vital detail which is lacking even in the Bible: the information given to Isaac by his father. As for Grossi, it seems rather odd that this occurs immediately after Abraham's conversation with the angel. The words 'Care fili Isaac . . . propera, defer huc ligna in Dei obsequium; volo sacratura dare, libare modo victimam . . .' are intoned in the middle of a musical phrase, a procedure deceptively suggesting the son's presence during the summons. The representation of non-consecutive events was obviously not the composer's forte (cf. his Esther dialogue, discussed in Chapter 2).

Cossoni, on the other hand, once more gives an example of his realistic interpretation of the story. Basing himself on the scriptural indication of the time ('media nocte'), he logically enough assumes that Isaac is sleeping and needs to be awakened by his father. Hence he interpolates a scene in which Abraham calls his son with increasing intensity (Ex. 84). What enhances the realism of this scene is the insertion of a general pause after each of the calls. One sees Abraham listening and waiting for a sound in the stillness of the night. When at last Isaac awakes from his sleep and is told that he has to accompany his father to a sacrificial service ordered by God, he gets up quickly, shouting 'venio'.

According to the Bible, the journey takes no fewer than three days. The problem of how to bridge this time in a dialogued representation of the story is solved (or remains unsolved) by each of the composers in a different way:

Capello:	A ritornello played by a treble stringed instrument with basso continuo.
Antonelli:	A *symphonia* based partly on the melodic material of opening piece.
Carissimi:	The Historicus tells in few words that father and son travel to the place indicated by God (recitative).
Grossi:	The time between departure and arrival is skipped.
Charpentier:	As with Carissimi, but strictly in accordance with the scriptural text.
Cossoni:	A composite duet for father and son, set to a free text.

Ex. 84

Capello's ritornello is a periodically conceived piece, spanning sixteen bars. Its joyful dance-like character may seem to refer to Isaac who, being unaware of what is in store for him, is delighted to accompany his father. Yet further on in the dialogue the same music is heard in quite a different context. Antonelli's *symphonia* does not depict anything specific; it simply bridges over the time. Both Carissimi and Charpentier make use of their Historicus, a convenient and legitimate means which temporarily turns the dialogue into mere narration. Grossi, not surprisingly, ignores the problem. Alone, Cossoni accepts the challenge to represent the journey by direct speech. While it is true that the first part of his duet is set to rather naïve words ('Eamus, festinemus per silentia noctis. Ad montem eamus quo vocat Deus; per tenebras ad montem eamus'), the lively music ('svelto' and 'presto') with melismatic motifs in imitation convey the impression of bodily movement. Then the duet proper is interrupted by a few phrases sung in recitative style:

ABRAHAM. Look, son, at the stars on the cloudless sky. So many descendants from our blood hath God granted us to be born.
ISAAC. I adore the Ruler of the universe, who governs everything with fixed laws.

The father's words are puzzling. It is true that before Isaac's birth God had already pledged himself in this sense (Gen. 15: 5). Yet the quotation of the divine promise at this very moment seems utterly ineffective in terms of drama, as it weakens the outcome of the story (the angel's statement made in verses 17–18). In my article dealing with Cossoni's 'Einsiedeln' dialogues I spoke, perhaps with some exaggeration, of a piece of clumsiness on the part of the author of the text.[4] Since then I have changed my opinion and can now offer an interpretation of the words. Abraham's absolute confidence in God induces him to tell his son about the promise, believing that eventually everything will be all right. Obviously he still underestimates the rigour of the charge. The joyful remainder of the duet set in 6/8 time to the text 'Iam mons apparet, quo vocat Deus, iam surgit aurora, eamus' also points to this. From these last words it can be deduced that, in Cossoni's version of the story, the journey takes only half a night instead of three days.

The short interruption of the journey on the third day (v. 5) is skipped by Grossi, Carissimi, and of course, Cossoni. Abraham's instruction to the two young men is included not only in Charpentier's setting but also in those of Capello and Antonelli, both of whom omit so many other, more important, details. The fact that in the Bible the patriarch's order to the servants is rendered in direct speech may have been decisive in this connection.

The preparations for the last part of the journey (v. 6) are described by both Carissimi and Charpentier. Contrary to the scriptural reading, in the text of the first it is not Abraham but Isaac who carries the 'knife' and the 'fire'. The

---

[4] Noske, 'Sacred Music as Miniature Drama'.

Frenchman's score contains a puzzling heading to this narrative passage: it has to be sung by 'duo iuvenes'. These persons cannot be identified as the two young servants who have just been left behind, since they are unable to relate events which they have not witnessed. Therefore I assume that these 'iuvenes' represent the Historicus. This is all the more likely since they reappear in the same capacity further on in the score.

Isaac's question about the missing victim and Abraham's answer (vv. 7–8) are set by all six composers. Capello, who prefaces the words by a repeat of his *symphonia* suggesting the act of walking, renders the text in a rather neutral way; the same is true of Antonelli and Grossi. Charpentier, on the other hand, extends the conversation to produce a particularly emotional verbal interchange between father and son. This is also the case with Carissimi; both he and Cossoni represent the scene only after the arrival at the place of the sacrifice. It is for this reason that the last three settings will be discussed below.

Sacrifice and Divine Intervention (Gen. 22: 9–14)

We now approach the climax of the drama. The practice of setting only those scriptural passages which are in direct speech, maintained both by Capello and Antonelli, proves particularly unsatisfactory here. Since in the Bible nothing more is spoken, either by the father or the son, Capello merely bridges over the time with a repeat of his ritornello, the dance-like character of which contradicts the gravity of the situation. Antonelli's procedure is even more awkward: Abraham's evasive answer 'My son, God will provide himself a lamb for the burnt offering' is followed immediately by the intervention from heaven. Both these composers dispense with the enaction of the sacrifice proper.

Unlike the dialogues by Capello and Antonelli, the four other settings do represent the scene, not only with biblical or freely invented description but in addition with highly emotional dialogue or soliloquy. As appears from the text of Carissimi's Historicus, Isaac's question about the missing victim disturbs the father deeply: 'Tunc obruit dolor patris viscera, fremuit sanguis, horruit natura.' These words are set in G minor, strongly contrasting with the previous major mode (D). The narrator's melody contains descending leaps of diminished intervals on the words 'dolor' and 'horruit', while 'fremuit' is depicted twice by a melisma of demisemiquavers. The ensuing dialogue between father and son is a particularly famous passage that has been reproduced in several modern publications (most recently in Dixon's monograph).[5] The composer expresses emotion not only by a continuous use of diminished intervals but in addition by a device which goes beyond the mere application of the rhetorical figure of *suspiratio*: words and even syllables are separated by rests (rendered below by ellipses):

---

[5] Dixon, *Carissimi*, 36.

ABRAHAM. Fili mi . . . heu . . . fili mi . . .

ISAAC. Pater mi . . . pater mi . . . quid . . . suspiras?

ABRAHAM. Heu . . . fili mi . . .

ISAAC. Pater mi . . . pater mi . . . quid suspiras? . . . Pater mi . . . pater mi, ubi est holocausti victima?

ABRAHAM. Providebit . . . Dominus . . . holocau . . . sti . . . victimam.

The Historicus, describing Abraham's final preparations for the sacrifice, keeps up the tension, as in his depiction of the binding of Isaac's hands (Ex. 85).

Ex. 85

Charpentier's version of the scene shows his familiarity with the work of his former master. He, too, changes the tonality from D major to G minor and encloses Isaac's question about the missing victim in the following extended piece of emotionally charged conversation (original or slightly amended biblical words in italics):

ISAAC. *Pater mi*, pater mi.

ABRAHAM. Fili mi, *quid vis?*

ISAAC. Quo properamus?

ABRAHAM. Hei mihi, fili dilecte!

ISAAC. Pater mi, quid suspiras? Cur non respondes? Dic mihi, quo properamus?

ABRAHAM. Sequere me, fili dilecte, sequere me. Parendum est Deo et sacrificandum.

ISAAC. Pater mi . . .

ABRAHAM. Fili mi . . .

ISAAC. *Ecce lignum, ignis* et gladius, sed *ubi est holocausti victima?*

ABRAHAM. Hei mihi, fili dilecte!

ISAAC. Pater mi, quid suspiras? Cur non respondes? Dic mihi, ubi est holocausti victima?

ABRAHAM. Sequere me, fili dilecte, sequere me. *Providebit sibi Deus holocausti victimam.*

The insertion of the words 'quo properamus' and 'sequere me' is explained by the fact that in this instance the scene is enacted *before* the arrival at the place of the sacrifice. Charpentier's setting is subtler than that of Carissimi. The composer builds up the tension gradually by the use of ascending chromaticism in the part of Abraham and its continuo bass. When, towards the end, both voices are combined, the father's exclamation 'Hei!' is conveyed by a suspended ninth resolving to the fifth below (Ex. 86). After a short rest ('faites icy une petite

Ex. 86

pause') the final preparations for the sacrifice are described in a largely homo-phonic chorus, parts of which are accompanied by doubling instruments. Charpentier discreetly depicts the biblical words 'desuper ligna comosito' with a diatonically ascending scale of a seventh. Particularly moving is the setting of 'et filium obedientem nec contra Deum murmuranem posuit super struem ligno-rum', alternately sung by two high and two low solo voices. During the final musical phrase pathetic suspensions of the ninth underscore the text: 'Tunc extendens manum arripuit gladium ut immolaret filium suum unigenitum.'

Totally different is Cossoni's version of the scene. Instead of answering his son's question evasively, Abraham bluffly replies: 'Thou, O my son, thou art the victim.' Still more astonishing is Isaac's reaction to these words: 'I shall die with pleasure. Let all my blood be drawn, if it pleaseth God.' This passage is sung in fast triple time denoting joy. To a modern mind, Isaac's prompt acceptance of his fate may seem at variance with psychological credibility. Theologically, however, it is entirely correct. The son, who has been informed about the divine com-mand (cf. Abraham's words earlier in the dialogue: 'sic iubet Deus'), cannot afford to oppose God, as in the course of time he will become a patriarch like his father. It is obvious that on the part of the author/composer this consideration prevailed over a 'human' representation of the boy's state of mind. Abraham's ensuing instructions to his son are given with kind words, yet without the sor-rowful sentiments shown in Carissimi's and Charpentier's settings: 'Bend thy head, close thine eyes, beloved son, do not fear the knife.' The victim's attitude remains submissive. His words 'I am waiting for death' contain a *topos*, also found in contemporary settings of the Credo in the Mass: a sustained note on the second syllable of 'expecto'. Abraham's emotion is 'postponed' until the very moment that he is about to kill his son. Here the textual procedure of having the patriarch speak to himself proves particularly effective: 'My trembling hand, why dost thou falter? Yield, yield paternal piety!' The principal key of C major, which despite occasional visits to neighbouring keys has been maintained during the course of the piece, now changes suddenly into that of E major. The chord of B major following D major underlines the fact that Abraham is trying to nerve himself, addressing God: 'Creator of heaven and earth, accept this holocaust of my blood.' The music, however, indicates that he has not yet overcome his weakness, the phrase ending in the key of E. But then an abrupt return to C major announces the awful act: 'Tibi haec victima cadit' (For thee this victim falleth).

As is shown by Grossi's text, a different representation of the scene is possible. Despite Abraham's words in the first part of this dialogue ('defer huc ligna *in Dei obsequium*' (bring the wood in obedience to God)), Isaac does not seem in this case to be aware of the fact that the sacrifice is at the command of divine authority. Hence his bewilderment at his father's operations: 'Why dost thou cover my eyes, why dost thou bind my arms?' (the non-biblical detail of Abraham covering the boy's eyes is also seen on Rembrandt's famous painting at the Hermitage in St Petersburg as well as on the beautiful etching of 1655). Isaac continues with affecting words: 'Why dost thou seize of me, what is thy purpose? Dost thou not know that I am thy Isaac? Is it I whom thou markest out for a miserable fate? What have I done unto thee, dear father? Art thou angry with me? Is it thus that thou lovest thy son? Give heed!'

Disturbed by these words, Abraham tries to defend himself but is interrupted by Isaac. The following phrases are for the greater part sung simultaneously with frequent repetitions:

ABRAHAM. Ah, my dear son, I know thee, thou art my heart, my life, my love!
ISAAC. No, I am not so, I am guilty of something.

Then Abraham gives the reason for his operations, explaining the situation: 'Be silent, my son, and listen to me. Heaven calls thee down as the victim. Thou art given up to death as a sacrifice for my sake. The command issues from the highest divine power. This is its will: that thou shalt die.' From here on Isaac acts in accordance with his future status of a patriarch: 'Proceed, proceed, my father. If it be the will of God, I do not ask further, I shall not mourn. I am pleased to die. Behold my head, my neck, give up thy son.' Subsequently, Abraham and Isaac simultaneously sing the following words: 'Let us offer a double gift to the Highest: I/thou my/thy blood, thou/I thy/my love.'

The transformation of the biblical story into a dialogue entails the turning of the divine intervention into a real *coup de théâtre*. Curiously enough, most of the works under discussion fail to take advantage of this opportunity. In Capello's and Antonelli's compositions the effect is practically non-existent, because of the missing description or scenic representation of the sacrificial act. Both Carissimi and Charpentier follow the scriptural text, with the result that the Historicus discloses the identity of the character who is going to speak. This considerably weakens the surprise of hearing the voice from heaven. While it is true that Charpentier prepares the *coup* by inserting, after the final words of verse 10 ('and took the knife to slay his only son'), a general pause of one bar, the continued use of the choir for the opening words of verse 11 ('And the angel of the Lord called unto him from heaven, and said') deprives the call 'Abraham, Abraham!' of its theatrical effect. Grossi and Cossoni, both of whom wrote their works entirely in direct speech, making them less dependent on the biblical reading, did not encounter this impeding factor. Yet Grossi's setting fails to exploit the scenic possibilities. With him, the angel's call occurs in the middle of a musical phrase sung simultaneously by father and son (Abraham: 'iam ferrum perstringo . . .'; Isaac: 'caede, caede natum . . .'). This deceptively conveys the impression that the divine words are linked with the foregoing dialogue instead of drastically altering the situation. Cossoni, on the other hand, shows his genuine dramatic talent by applying a very original procedure which not only does justice to the scene as such but, moreover, adds a dimension to the divine intervention (Ex. 87).

Ex. 87

Abraham's final word 'cadit' is depicted by a downward scale spanning no less than a twelfth. This is an instance of word-painting in the literal sense of the term; that is, only the *word* is painted, not the *act*. The act itself is checked by the following chord on the note E in the continuo. Hereupon, we hear the angel's voice: 'Hold! Put down thy sword.' The key offers a surprise: F major, appearing here for the first time in the composition. The preceding bar with the chord on E seems problematic, however. The question is: which chord exactly had Cossoni in mind?

Theoretically speaking there are four solutions. Two of these, a minor triad (E–G–B) and a sixth chord (E–G–C) can be dismissed; they are musically weak and quite ineffective in terms of drama. If, on the other hand, we assume that the note E is a scribal error and that instead the composer intended to write the low C, this would result in a perfect cadence in that key, an impeccable procedure. Yet such a cadence is at variance with the required dramatic tension, carefully prepared by the progressions C major–E major and D major–B major earlier in the scene. It seems most unlikely that for purely musical reasons Cossoni would have allowed such an anticlimax. Another argument against this solution is the fact that there are practically no scribal errors to be found in the autograph. The fourth possibility is a major triad on E. Although distinctly daring, this produces a splendid effect. The lack of a sharp above the note in the continuo part is not at all significant. Here, as elsewhere in the dialogue, the figuring of the bass is scant, and many useful indications (figures and/or accidentals) are absent. Therefore there can be no question but that the E major chord is the only acceptable solution.

The chord represents an act of God, the paralysing of Abraham's arm. Let us not forget that in a sacred dialogue the lack of visible action has to be compensated for by musical means. Without the chord, the angel's words would come too late, Isaac having already been struck by the sword. This is not to say that the text of the angel is superfluous; it conveys the indispensable rational information. But before that the intervention of God had to be expressed by music alone.

My interpretation of the E major chord as a symbol of the divine act is reinforced by a musical particularity. It is an isolated chord defying the laws of tonal progression. True, the sequence E–F can be explained theoretically as an interrupted cadence in A minor, whose submediant assumes the function of the tonic in F major. But this academic interpretation does not hold water. The chords of E and F are separated by a meaningful general pause of half a bar prolonged by a fermata, so there can be no question of a genuine cadence. The musical isolation of the E major chord, denoting divine absolutism, signals the intervention of Heaven.

Verse 13, describing the ram caught in a thicket and subsequently used for the sacrifice instead of Isaac, was set only by Capello and Charpentier. The latter assigned these words to the 'duo iuvenes', acting again as narrators. As for Capello, he finally departs from the exclusive use of biblical direct speech by setting the verse for all three voices with instruments and makes it the concluding section of the composition. This means that his work lacks an essential element of the story: God's benediction of the patriarch and his posterity. None of the composers sets verse 14, dealing with the name-giving of the place of sacrifice. The omission is hardly surprising, as this detail adds nothing to the essence of the story. In its place Carissimi interpolated in his dialogue a small duet, set to freely invented words, which express the joy of father and son. Its formal scheme is that of a da capo structure; this involves the recapitulation of the opening phrase starting with the repeated cry 'O!' Since exclamations lose their spontaneity by distant restatement, the effect is somewhat unsatisfactory. The duet's middle section ('Away with the fire, away with the sword, away with death!') contains a meaningful turn to the minor mode on the word 'mors'.

Benediction (Gen. 22: 15–18) and Conclusion

In the Bible the angel calls Abraham a second time, quoting God's benediction of the patriarch and his progeny. Only Carissimi keeps to the scriptural reading, setting it first almost literally and subsequently in a paraphrased form. This is followed by a *conclusio* addressed to the listeners. Antonelli, Grossi, Cossoni, and Charpentier amalgamate the two calls, the intervention being immediately followed by the benediction. All these composers use abbreviated scriptural texts, scarcely paraphrased by Antonelli, Cossoni, and Charpentier, but treated much more freely by Grossi. Grossi does not even mention Abraham's posterity explicitly; there is only a vague allusion to it ('coronata pietas'). Cossoni has a short concluding section a 3, addressed to the congregation. Here the individual characters lose their separate identities: 'Sic coronatur qui vincit se. Justus es, Domine. Non confundetur qui credit, qui sperat, qui diligit te.' The final words are borrowed freely from St Paul (1 Cor. 13: 13).

Antonelli's concluding double choir with instruments is set to the condensed verses 18 and 17 (in that order). This impressive piece of music, which combines various techniques, such as homophony, trio settings for two sopranos, altos or tenors with continuo, and antiphonal treatment of semichoruses, demonstrates the composer's remarkable proficiency. Grossi structures his 'finale' as a continuous alternation of virtuosic solo phrases, the melodic material of which he ingeniously uses for a concluding terzet of eleven bars. Cossoni's ensemble is written in transparent counterpoint, interrupted by the homophonic setting of the words 'Justus es, Domine'. Because of their brevity, his final statements leave the effect of the above-mentioned climactic setting of the intervention from heaven unimpaired.

Carissimi's slightly shortened and partly paraphrased setting of verses 15–18 is sung successively by the Historicus and the angel, the latter's recitative tending to arioso style. The ensuing quintet, a praise of God alluding to Psalm 116, shows a relationship, both textually and musically, to the solo of the angel. The motifs previously set to the words 'omnes populi' and 'omnes gentes' now function as the structural material of the ensemble. As in Cossoni's piece, the singers abandon here their separate roles. The modern edition included in the *Opere complete* suggests that they resume them during an intermediate section set to the text '[Et adorate Dominum], qui misit Angelum suum de coelo, et eripuit Isaac dilectum de igne, de gladio, de manu patri sui.' This, however, makes no sense; it would mean that God, singing the bass, praises himself. The quintet forms a much-needed polyphonic conclusion of Carissimi's dialogue, which, apart from the little duet of father and son, consists almost entirely of sections set in recitative or semi-recitative style.

Much more complex is the way in which Charpentier concludes his version of the drama. While the reduction of the two calls to a single one entails a textual amalgamation of intervention and benediction, the two statements are expressed musically in different styles. Verse 12 is set almost entirely as a recitative; the abbreviated verses 17 and 18, on the other hand, are sung in aria style. The lively triple-time melody of the latter subsection, including melismas supported by a 'walking' bass, adds to the 'dry' quotation of God's words the expression of the angel's individual sentiments. In the score this is alluded to by the tempo indication 'guay', a literal translation of 'allegro' in its original meaning.

After an instrumental ritornello of nine bars, the intervening verse 14, dealing with the ram caught in a thicket, is set for the two high voices of the 'duo iuvenes' representing the Historicus. Subsequently Abraham and Isaac sing a duet: 'Gaude pater fidelissime/Gaude fili obientissime'; this freely invented text once more points to Charpentier's familiarity with the work of Carissimi. Like the latter's duet, discussed in the previous section of this chapter, it includes the words 'away with the fire' and 'away with the sword'.

Unlike Cossoni and Carissimi, Charpentier maintains the identity of his characters during the greater part of his 'chorus ultimus'. This is a particularly complex piece, offering a great variety of scoring. Ensembles of solo voices alternate with the full choir, which in turn sings alternating passages with and without doubling instruments. In addition, two violins twice interrupt the text with short interludes. The opening words of the Historicus, set for three solo voices, quote the scriptural verse 19: '[Et] reversus est Abraham ad pueros suos;' for the remainder of the text the singer adopts the role of a non-biblical narrator, announcing a song of thanksgiving by Abraham and Isaac. This entails the insertion of a second duet. It starts with the words 'Beati qui timent Dominum', alluding to Psalm 111: 1, and then mentions (at last!) the divine benediction in 'guay' tempo: 'quoniam augebit Deus et multiplicabit semen eorum in saeculum'. It is only at the moment when the full choir repeats these words that the characters abandon their roles; they now clearly address the congregation.

The extensive 'chorus ultimus' is in perfect balance with the dimensions of the work as a whole. As regards musical quality, it demonstrates not only Charpentier's extraordinary compositional abilities but also his exceptional diversity of expression.

Evaluation

It is a well-known fact that, in matters of aesthetics, evaluation and scholarship are fundamentally at variance. Yet this is merely a theoretical observation. In practice no scholar discussing works of art can avoid making a value-judgement of his material. A means of 'objectifying', to some extent, his personal opinion is its underpinning by falsifiable arguments. On the basis of these, it will be possible to accept or reject the scholar's judgement and to arrive, perhaps 'inter-subjectively', at a *communis opinio*. The following comparative evaluation of the Abraham dialogues will be made from distinct points of view: musical and dramatic. It is obvious that these two avenues of approach can lead to different conclusions. Whereas a highly rated piece of music may lack dramatic qualities, it sometimes happens that a mediocre composition turns out to be particularly effective in terms of drama.

Regarding our first point of view, that of the musical character, the dissimilarity of the six works precludes the application of fixed criteria to all of them. It would indeed be grotesque to compare in this way Capello's modest three-part piece, published in 1615, with Charpentier's version, dating from the 1680s and set for eight solo voices with a four-part choir. Therefore I shall examine the compositions in pairs: that is, Capello–Antonelli, Grossi–Cossoni, and Carissimi–Charpentier. This grouping is anything but arbitrary. Both the works of Capello and Antonelli are of restricted dimensions and the difference of scoring between them is apparent only in Antonelli's eight-part concluding section, the textual content of which is lacking in Capello's setting. Cossoni presumably wrote his *Sagrificio d'Abramo* about a quarter of a century after Grossi's dialogue. Yet the two Lombardic composers were born only eight years apart (1623 and 1615 respectively), and their musical styles have much in common. Moreover, both show a preference for the treatment of biblical subjects with non-biblical words, a practice that enhances the freedom of their musical expression. Charpentier's *Sacrificium Abrahae*, too, was written much later than Carissimi's *Historia*. In this instance, it is the master–pupil relationship which, far from being a mere historical circumstance, proves particularly important. Unlike Grossi and Cossoni, these two composers made ample use of the scriptural text. The fact that Carissimi's piece is an oratorio dialogue, while Charpentier's is not, must be considered immaterial in this context; it in no way affects the stylistic affinities between the two works.

Both the early pieces show an expert handling of melody. Capello's setting, more closely connected with the text than Antonelli's, contains beautiful melodic contours, such as the ascending line illustrating the words 'super unum montium' (v. 2) and the leap of a seventh accompanying Isaac's question 'ubi est victima holocausti?' (v. 7). Simultaneous use of voices and instruments occurs only in the concluding narrative section (v. 13). Antonelli's abstract treatment of the text has a more colourful background, the instruments accompanying the vocal parts during the summons of Abraham (vv. 1–2) and the intervention by the angel (vv. 11–12). However, the main qualitative difference between the two settings is in their structural proportions. Capello's dialogue is a much more balanced piece of music than that of Antonelli. The former's restatements, both of instrumental sections (*symphonia*, ritornello) and vocal passages (the call 'Abraham, Abraham' with the answer 'Adsum'), contribute to the overall equilibrium. Antonelli, on the other hand, rounds off his work with an eight-part chorus, the proportions of which are at variance with the brief and rather fragmentary setting of the dialogue proper. The high quality of this ensemble does not compensate for the structural weakness of the composition as a whole.

Despite the affinity between the musical idioms of Grossi and Cossoni, their settings are dissimilar. Grossi exploits his virtuosic compositional abilities. Recitative, semi-recitative, arioso, and aria (as style, not as form) continuously alternate without breaks, a procedure which is sometimes at variance with the piece's verbal structure. It is true that on several occasions the melody expresses or depicts the text with great accuracy. Elsewhere, however, words are gratuitously repeated for the sole reason that a certain motif demands restatement. As for harmony, the sequence of tonalities seems arbitrary; the composer wanders aimlessly around the main key (C minor). This proto-tonal indeterminacy is entirely justified during the heated conversation between father and son, preceding the latter's acceptance of his fate. However, on other occasions, such as the information given to Isaac during the early phase of the story and the crucial moment of the intervention from heaven, it proves unsatisfactory. All this does not mean that Grossi's composition must be given a low rating—on the contrary, it contains delightful music. Yet its value lies in the details rather than the overall structure. The lack of planning suggests almost that the piece was written in a hurry.

Unlike Grossi, Cossoni must have planned his work carefully. His ingenious handling of tonalities during the drama's climax has already been discussed. This was my main reason for guessing the date of composition as late as the 1680s. While Grossi completely dispensed with a divisional layout, Cossoni treats each of the story's episodes separately, rendering them in musical sections which close with perfect cadences often followed by a rest. Since this procedure may result in an unsatisfying mosaic-like structure, it is compensated by frequent contrasts of tempo, which, though occurring at unexpected moments in the musical discourse and sometimes overlapping the sections, are textually always justified. Another means of avoiding a static representation of the episodes and promoting continuity is the evolution of semi-recitative into arioso within the separate sections. A particularly happy contrivance is the insertion of the duet depicting the journey. This 'piece' considerably enlivens the work as a whole. The animated treatment of the continuo part, which occasionally shares in the voices' melodic material also contributes to the overall variety of expression. Being strongly dependent on its text, Cossoni's composition lacks the abstract virtuosity found in the dialogue by Grossi. Nevertheless, it is an attractive piece of music.

An essentially different musical conception appears in Carissimi's *Historia*. If Cossoni depends on his text, we must realize that this is for the greater part a freely dialogued text, fashioned for no other purpose than to be set to music. Carissimi, on the other hand, bases his setting on the biblical reading of the story, that is, a text predominantly written in narrative form and not intended to be sung. In consequence, a part had to be created for the Historicus, whose neutral position in the drama did not allow any other means of musical expression than recitative. The assignment of this neutral role to a single performer—a logical procedure adopted by the composer—may lead to monotony. While it is true that Carissimi was particularly expert in the rendering of affections, both in monody and ensembles,[6] only a few passages in the scriptural narration afford an opportunity for writing expressive music. This may explain the abbreviation and paraphrasing of the biblical words as well as the omission of verses 4–6 and 13–14. Nevertheless, the first part of the *Historia*, mainly transmitted by the narrator, makes a rather reserved impression. It is only when the composer arrives at an episode which is enriched with freely invented words that the musical language becomes fascinating: the emotional conversation between father and son described earlier in this chapter. During the remainder of the composition the variety of expression is maintained. The angel's message, for instance, following the inserted duet of Abraham and Isaac, evolves from recitative into semi-recitative; its cantabile melody could even be termed 'arioso'. The joyful concluding quintet strongly contrasts with the work's 'dry' opening.

Charpentier's composition has the character almost of a stylistic elaboration of Carissimi's *Historia*. Everything contained in the latter's work is offered here on a larger scale. The narrator not only speaks through the mouth of a single performer but is represented in addition by various ensembles: duets, terzets, and four-part choruses. The resulting variety of scoring removes the necessity for textual abbreviation; of the six composers, Charpentier makes the most ample use of the scriptural reading, albeit often with slightly changed words. Instead of one inserted duet there are three: Sarah–Abraham and twice Isaac–Abraham. The emotional scene between the father and son is more extensively composed than the corresponding one in Carissimi's dialogue; the same is true of the complex 'chorus ultimus'. The partly unspecified instruments doubling portions of the narrative choruses and the *conclusio* add colour to the voices. This seems to be the only specifically French aspect of the work. Otherwise, the musical language is Italianate, representing particularly in the duets the north Italian style rather than that of Rome. Little needs to be said about the overall quality of the composition. Charpentier fully displays his extraordinary skill in all the techniques used. The most remarkable among these, applied in small-voiced ensembles as well as the concluding chorus, is brilliant counterpoint. Though unacademic, this counterpoint may nevertheless be called 'Classical'. While several previously discussed settings of the story evince arresting characteristics, in terms of pure music none of them can match Charpentier's *Sacrificium*.

A value-judgement of the dramatic qualities, unlike that of the music, allows in principle a comparison between all six compositions. Yet in this field of enquiry,

---

[6] '[Carissimi] surpasses all others in moving the minds of the listeners to whatever affection he wishes' (A. Kircher, *Musurgia universalis* (2 vols.; Rome, 1650; repr. 1969), i. 603).

too, the dialogues by Capello and Antonelli stand apart from the other works. The reason for this is not their early date but the fact that their authors/composers completely disregarded certain important events and circumstances related in the Scriptures, such as the act of sacrifice and the divine benediction (the latter omitted only by Capello). Actually, compared to the biblical story, both texts make a rather incoherent impression, and it is difficult to accept that the scores that have come down to us represent exactly the way in which they were performed in church. A solution of this problem can only be conjectural. Hence I advance with some hesitation the hypothesis that the missing items were inserted or added, either in spoken form or in chant. Presented in this way, the dialogues would be in agreement with the biblical text as a whole. The same practice might have been adopted in the Crocifisso oratory, assuming that Antonelli wrote his piece for that institution.

However this may be, the fact remains that both works ignore almost entirely the sentiments of the characters involved, in particular those of the father. As a result of this, an essential element of drama is missing: conflict. Neither the texts nor the musical settings allude to the struggle between paternal piety and obedience to God. And so we must conclude that in these instances a subject loaded with dramatic implications has been realized in an undramatic form. Although Capello and Antonelli restricted themselves mainly to the rendering of those biblical text portions that were written as *oratio recta*, paradoxically their representation of the story comes across as narrative rather than dramatic.

As for the four other works, several devices applied in one or more of them were probably founded on theatrical tradition. Théodore de Bèze's 'tragédie' *Abraham sacrifiant* (1550) contains a 'cantique d'Abraham et de Sara' praising God, which is paralleled by the duet in the introductory part of Charpentier's setting. Another, more important precedent encountered in this Calvinist play is the transposition of Isaac's question about the missing victim to the moment of arrival at the place of the sacrifice (Carissimi, Grossi, Cossoni). It is easy to see that this procedure serves the dramatic efficacy. Moreover, as in Grossi's piece, De Bèze's Isaac does not readily accept his fate. He asks for pity:

> Helas, pere tresdoulx,
> Je vous supply, mon pere, à deux genoux,
> Avoir au moins pitié de ma jeunesse.

Although there is no question of a conflict between father and son, the French text shows some affinity with that of Grossi:

Grossi	De Bèze
Nescis me tuum Isaac? Me miserum a re destinas? Quid, quid feci tibi, care pater?	Mais, mais voyez, ô mon pere, mes larmes, Avoir ne puis ny ne veulx autres armes Encontre vous: je suis Isac, mon pere, Je suis Isac, le seul filz de ma mere: Je suis Isac, qui tien de vous la vie: Souffrirez vous que'elle me soit ravie?

Still another parallel is found in the following lines, corresponding with the words 'sum reus' (I am guilty) in Grossi's dialogue. They are addressed to God:

> Seigneur, tu m'as cree et forgé,
> Tu m'as, Seigneur, sur la terre logé,
> Tu m'as donné ta saincte cognoissance,
> Mais je ne t'ay porté obéissance
> Telle, Seigneur, que porter je devois.
> Ce que je prie, helas, a haulte voix,
> Me pardonner.

One could question, of course, the likelihood of Grossi's borrowing from a play written more than a century before by a Protestant humanist. However, the relationship between the two texts can be explained otherwise, namely by De Bèze's dependence on contemporary Latin Catholic sources, such as the plays by Hieronymus Ziegler (*Isaaci immolatio*, dating from 1543 and printed five years later in Basle), and Philicinus (*Dialogus de Isaaci immolatione* (Antwerp, 1546)).[7] Therefore, it cannot be excluded that the handling of the details quoted above belonged to a tradition still alive during the seventeenth century.

Grossi's conception of the emotion-laden scene at the place of the sacrifice is dramatically arresting, in regard to both text and music. Unfortunately, this is not the case with other events in his dialogue. The fact that Isaac and the angel have the same vocal tessitura (soprano) already weakens the 'theatrical' effect of the latter's intervention. This event in the story (as well as others) remains unhighlighted because of the composer's excessive striving for musical continuity. Other examples include the information given to Isaac about the purpose of the journey, the time needed for reaching the mountain indicated by God, and the demarcation of the *conclusio* from the dialogue proper (it is not entirely clear whether the characters here step out of their roles or not).

Unlike Grossi, Carissimi builds up the tension gradually; it reaches its climax in the scene preceding the divine intervention. Yet his text, heavily depending on the scriptural source, precludes full exploitation of the story's dramatic possibilities. Apart from the narrative character of the words sung by the historicus, the representation of the intervention and the benediction, separated by a static duet, affects the impact negatively. Similar defects are apparent in Charpentier's work which, by virtue of its greater dimensions, both in scoring and length, requires stronger dramatic effects. The main weakness in the textual layout stems from three unhappy decisions: (*a*) the retention of Isaac's question about the missing victim at its original place, that is, during the journey and not after the arrival; (*b*) the superfluous description of the substitute offering of the ram, causing a break in the continuity; (*c*) the postponement of the praise of God by

---

[7] See the editorial introduction to Théodore de Bèze, *Abraham sacrifiant*, ed. K. Cameron, K. M. Hall, and F. Higman (Geneva and Paris, 1967).

Abraham and Isaac until after their return to the two young domestics. In all these three cases the effect is that of retardation resulting in a temporal distance between occurrences that, in terms of dramatic efficiency, should be represented consecutively. Isaac's question becomes divorced from its consequence, the act of the sacrifice, and the two duets of father and son, referring to the preservation of Isaac's life and the benediction of Abraham's posterity respectively, are deprived of their spontaneity. Taking into account Charpentier's familiarity with the stage, it is most unlikely that he had anything to do with the writing of the text. Rather, it seems that in this instance he became the victim of his librettist's inexperience.

If Charpentier's libretto reveals flaws, that of Cossoni is almost perfect. It is true that Isaac's prompt acceptance of his death 'with joy' challenges our credibility but, as has been said before, theological considerations prevailed here over psychological realism. Otherwise the text, omitting all superfluous details, stresses only the essential parts of the story. This leaves room for the insertion of freely invented meaningful scenes, such as the wakening of Isaac and the statements made by father and son half-way through the journey. Right after the summons the tension underlying the words becomes palpable and remains so during the journey and the conversation about the missing victim. Yet this tension is felt by the listener rather than the protagonist, whose bluff answer to Isaac's question ('*You* are the victim')—deviant from the Bible—is given almost automatically. It is only at the very last moment that Abraham faces reality and becomes aware of his weakness. The fact that he expresses his feelings in a soliloquy is a particularly original feature of this scene; it enables a freedom of expression that would be unattainable in a dialogued form.

More than the other settings, that of Cossoni resists the differentiation between musical and dramatic characteristics; the two are continuously integrated. An example is the depiction of the journey. From a musical point of view, the insertion of a duet contrasting with the previous use of semi-recitative is most satisfactory. However, at the same time both the text and the music of the duet convey dramatically to the listener the idea of travelling to a (seemingly) fatal destination. It is indeed remarkable that, among the six dialogues, this one, written by a 'minor' composer, offers the most homogeneous and convincing dramatic representation of the story.

APPENDIX

Performance Practice: The
Visual Aspect

The sight of the gestures and movements of the various parts of the body producing the music is fundamentally necessary if it is to be grasped in all its fullness.

Igor Stravinsky[1]

A few years ago a concert took place in The Hague, the programme of which included among other seventeenth-century Dutch works Verrijt's dialogue 'Fili, ego Salomon', discussed in Chapters 1 and 4 of this study. Everything seemed to point to an excellent performance. The three singers, a powerful bass and two high tenors, were not only equipped with beautiful voices; they also had considerable experience in early Baroque vocal style. As for the audience, its interest in music prior to Bach and Handel could be taken for granted, and those who were unfamiliar with Latin could read the text in the programme booklet which contained both the original words and a Dutch translation. Yet, despite these ideal circumstances, the performance fell flat. True, we heard a highly attractive piece of music, exquisitely sung. But nothing was done to make it clear that something *happened* in this composition: a father admonishing his son to avoid bad company, the latter admitting with remorse that he had sinned, the father's subsequent advice to pray in order to obtain forgiveness, and the final warning to all young men. Instead, far from acting a role in this miniature drama, each of the performers sang his part dutifully, almost unaware of the character standing at his side. Needless to say, the ritornello-like interruptions of the dialogue proper (in which the characters, temporarily abandoning their roles, address the congregation) were hardly understood as such, and the same was true of the moralizing *conclusio*, likewise addressed to the audience. A sung conversation carrying a message was presented almost as if it were an abstract piece of music.

Everyone active or interested in the realm of early music will be aware of the difficulty or impossibility of transmitting its original social or religious function to a twentieth-century concert audience. Yet music composed in far-off days is not exclusively time-bound. A dialogue dating from the seventeenth century is still a dialogue today; hence a conversation represented by 'statues' instead of living bodies is absurd, regardless of the epoch in which it takes place. Why, then, are the inherent visible aspects of human interaction so often neglected in modern performances of unstaged musical scenes?

The main reason, I believe, is the mistaken 'veneration' of everything that is old. As a

---

[1] I. Stravinsky, *Chroniques de ma vie* (Paris, 1935–6; 2nd edn., 1962); Eng. trans. as *An Autobiography* (New York, 1962).

result of this the concert hall or the church in which the performance takes place becomes mentally transformed into a museum; the listeners 'stare' with their ears at the music, and there is no true communication between performers and audience. Endeavour to give the audience the opportunity to take part in the 'happening', using not only the human faculty of hearing but in addition that of sight, is considered unbecoming, particularly in the case of sacred music. The theatre, on the other hand, does not suffer from these inhibitions; it may even appeal to senses other than hearing and sight. When, during a Paris performance of Rameau's opera *Les Indes galantes* in the 1950s, the 'Ballet des fleurs' was underscored by a perfume diffused in the theatre, only a few puristic people were shocked; the effect was simply delightful.

In order to draw the audience into a musical scene presented in a concert hall or church, the performers need to act visibly to a certain extent. This implies that they should know their parts by heart. Instead of looking into the score they should look at each other, make appropriate gestures, and take small steps forwards or backwards in accordance with the relevant situation. There is no need for a producer; this would lead to over-acting. In general, the performing musician, especially the singer, is already quite familiar with acting; he or she plays the principal role in the rite of a public concert. Assuming that the singer has a thorough knowledge of the text and its visual implications—an essential condition—he should act not only with his voice but also with his limbs, yet more soberly than in the theatre. His sole aim is to make it understood that he is a character, not merely a performer.

To make my meaning clearer, I give the following description of the enactment of a biblical dialogue by Giovanni Antonio Grossi, a work discussed at length in Chapter 2 and transcribed in full in Part Two. This is 'Heu! Domine, respice et vide', dealing with the announcement of the birth of Isaac. Although there are five vocal parts, the number of roles is only three, since the angels invariably sing as a trio. As such they represent God; hence they are addressed both by Abraham and Sarah in the singular as 'Domine'. This is in strict accordance with the scriptural text (Gen. 18: 1–15).

Bars 1–22   (Abraham: 'Heu Domine . . .'). Abraham (right) and Sarah (left) are standing next to each other in the centre of the 'stage'. The angels are invisible. Addressing God, Abraham looks upwards during the beginning of his solo. A sober gesture accompanies his words 'Ecce vernaculus meus . . .'.

23–37   (Sarah: 'Spera, nec despera, o caro sponse . . .'). While singing, Sarah looks at her husband. A gesture underlines the words 'Spera igitur . . .' (bars 33 ff.)

38–50   (Abraham, Sarah: 'Igitur speramus . . .') Both singers look at each other. During this small duet the angels appear from the right, the moment depending on the distance they have to cover. They approach slowly.

51–7   (Abraham: 'Ecce video de longè . . .') During the general pause preceding this recitative Abraham turns his head to the right and points to the angels. Sarah steps aside but remains visible.

57–69   (Abraham: 'Domine, si inveni gratiam . . .') The angels having reached the centre-stage, the patriarch looks them straight in the face.

70–99   (Angels: 'Fac. Ecce adsumus . . .') The angels address Abraham with small gestures. During their terzet Abraham turns once or twice to the left with a gesture suggesting his giving orders to his servants. Meanwhile Sarah takes a step back, making it clear that her husband is addressing not her but the invisible servants.

100–6   (Abraham: 'Ecce parata sunt omnia . . .') Abraham turns to the angels, accompanying his words by a gesture.

106–18   (Angels: 'Edamus ergo, socii . . .') During this part of their terzet the angels look at each other.

118–28   (Angels: 'Indica nobis, o care hospes . . .') The angels turn to Abraham.

129–39   (Abraham: 'Ecce in tabernaculo . . .') Abraham points to Sarah.

139–56   (Angels: 'Revertens veniam ad te . . .') The angels addressing Abraham adopt a solemn attitude. Sarah stares at them.

157–64   (Sarah: 'Oh, oh Deus meus . . .') During bar 156, prolonged with a fermata, Sarah takes a step forward, smiling. While she sings, the angels look at her.

165–74   (Angels: 'Quare risit Sara . . .') The angels address Abraham.

175–80   (Sarah: 'Non risi, Domine . . .') Sarah approaches the angels, while Abraham takes a step backwards. Sarah's face expresses fear until bar 212.

180–95   (Angels: 'Ridens dixisti . . .') The angels address Sarah in a severe tone.

196–212 (Sarah: 'Ignoscas, Domine . . .') Lively exchange of speech between Sarah and the angels with gestures on both sides. Standing back, Abraham does not take part in this episode.

212–20   (Sarah: 'Etiam, Domine, sed negavi . . .') Sarah adopts a submissive attitude, expressed by an appropriate gesture.

220–3   (Angels: 'Credite igitur . . .') These words are addressed to Sarah. Then Abraham takes a step forward, standing again between his wife and the angels.

223–43   (Angels: 'In vobis benedicentur . . .') Solemn statement of the angels without gestures. They address Abraham rather than Sarah.

243–58   (Abraham, Sarah: 'Credimus Domino. Gratias agimus . . .') Both address the angels with sober gestures.

259–75   (Angels: 'Ecce, quia credisti . . .') The angels again address Abraham.

275–80   (Abraham, Sarah: 'Gratias agimus Deo . . .') Abraham and Sarah express their gratitude with more vivacity than before. Since this is clearly apparent from the melismatic settings of their words, there is no need for additional gestures.

281–313 (Tutti: 'Gratias agimus Deo . . .') All five performers, stepping out of their roles, address the audience without gestures.

It goes without saying that the above description of the enactment is only tentative. Moreover, other subjects will require different kinds of solution. Yet, even in less spectacular scenes, like that of the frequently encountered conversation between an angel and a sinner, it should be made clear by visual means that it is a scene not merely a reflective or narrative text set to music. Despite the confusing seventeenth-century terminology, there is an essential difference between a motet and a dialogue.

PART TWO

Transcription of Ten Dialogues

PREFATORY NOTE

The compositions collected here are all appearing for the first time in a modern edition. They represent the dialogue genre in several of its aspects: its variety of subject (biblical or freely invented), scoring (two to nine voices with or without concertizing instruments), and style (north Italian, Roman, north European, and French). Since a critical edition of twenty-seven dialogues dating from the first three decades of the seventeenth century has already been published in 1985 (Smither *SDD*), with a single exception the selection in this edition favours works written after 1630. Each piece contains a reference to its description or discussion in Part One; the sources are given in the List of Dialogues. The editorial policy has been to correct tacitly errors in the original partbooks or scores and to add or adjust barlines. The white notation in the dialogues by Caruso and Grossi is reproduced in its original form.

CONTENTS

1. Heu mihi!

Dialogo per Alto, due Tenori e Basso *

[Peccator et tres angeli]

Alessandro Grandi

Il 2º libro de motetti (1613)

* See Part One, p. 115.

Alessandro Grandi

* This part makes itself invisible to answer [the first tenor] in echo.

Alessandro Grandi

* [The singer] reappears.

2. Ecce spina

Dialogo a 2, sopra l'aria della Pavaniglia, di Santa Rosalea *
Canto e tenore, con le riposte in eccho

Giuseppe Caruso
Sacri lodi (1634)

* See Part One, pp. 54 f.

Guiseppe Caruso

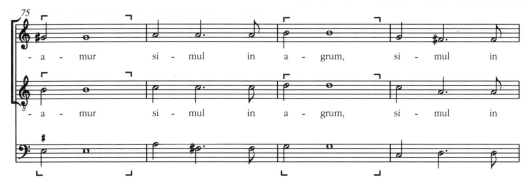

Guiseppe Caruso

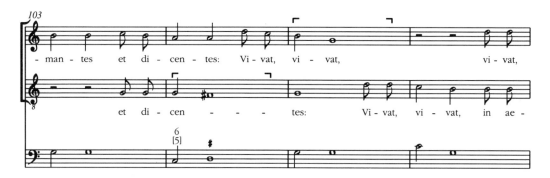

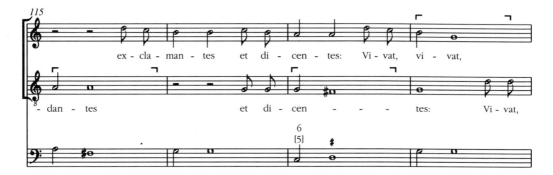

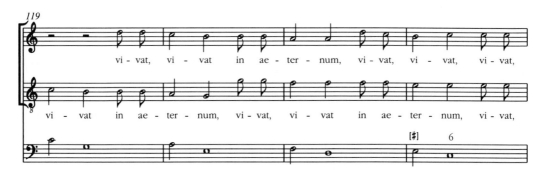

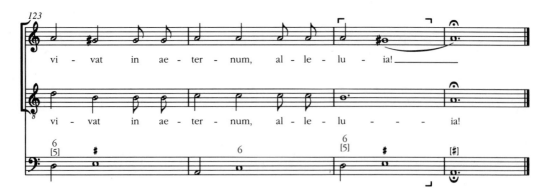

3. Gloria in altissimis Deo

Dialogo fra gli angeli e pastori nella Natività *

di nostro Signore

Chiara Margarita Cozzolani

Salmi ecc., op.[4] (1650)

* See Part One, pp. 28, 62.

4. Doleo et poenitet me

in Dialogo *

SSTB con 3 viole da gamba   [Duo peccatores, Christus, Deus Pater]

Giacomo Carissimi
MS., Uppsala, Univ. Libr.

* See Part One, pp. 120 f.

Lento
Deus Pater

Do - le - o,

do - le - o 　　et poe - ni - tet me, 　poe - ni - tet me, 　quod

Giacomo Carissimi

i - stos mi - ser - ri - mos pec-ca - to - res, qui - a non ces - sant cla - ma -

- re, cla - ma - re ad me.

Iam e - xau - di - vi e - os, iam,

5. Dialogo delle due Marie

a 2 voci *

Giovanni Legrenzi

Harmonia d'affetti devoti (1655)

* See Part One, p. 75.

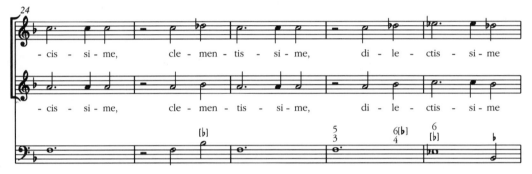

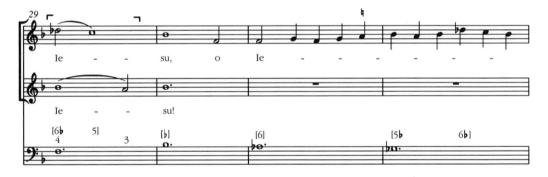

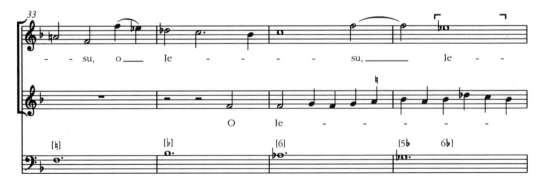

Ie - - su! Nul-lum bo-num si- ne te, om-ne ma-lum abs-que te.

Ie - - su! Im-mo si- ne

te om-ne bo-num est ma-lum, at-que te-cum om-ne ma-lum est bo - num.

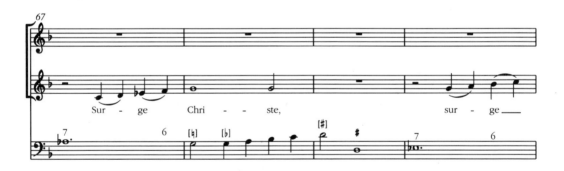

Sur - ge Chri - - ste, sur - ge

Re - di - me nos, li - be - ra

Chri - ste, ad - iu - va nos,

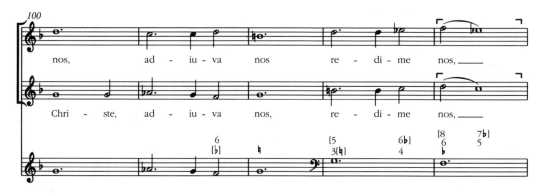

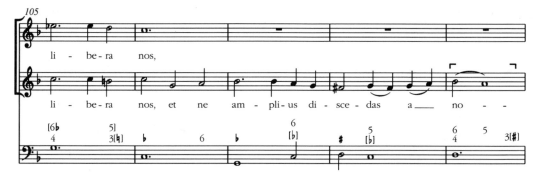

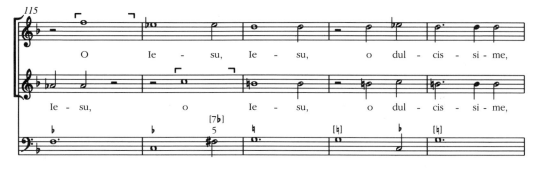

O Ie - su, Ie - su, o dul - cis - si - me,

Ie - su, o Ie - su, o dul - cis - si - me,

cle - men - tis - si - me, di - le - ctis - si - me Ie - -

cle - men - tis - si - me, di - le - ctis - si - me Ie - -

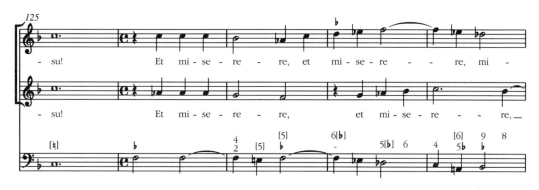

- su! Et mi - se - re - re, et mi - se - re - - re, mi -

- su! Et mi - se - re - re, et mi - se - re - - re, __

- se - re - re no - bis, mi - se - re - re ___ no - bis.

__ mi - se - re - re no - bis, et mi - se - re - re no - bis.

6. Heu! Domine, respice et vide

Dialogo: Abramo, Sara e tre Angeli *

Giov. Antonio Grossi
Ms. Milano, Archivio del Duomo

* See Part One, pp. 13 f., 17, 20–1, 36 f. and 197 f.

Giovanni Antonio Grossi

Giovanni Antonio Grossi

à 9 voci: Canto e due Cori *

Alessandro Della Ciaia
Sacri modulatus (1666)

* See Part One, p. 75.

Alessandro Della Ciaia

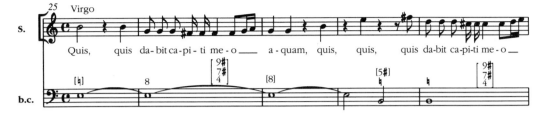

Quis, quis, quis da-bit ca-pi-ti me-o — a-quam, quis, quis, quis dabit ca-pi-ti me-o —

a-quam, et o-cu-lis me-is fon-tem la-cri-ma — — —

— rum? Et plo-ra — bo, — et plo-ra — bo te

De-um me-um, fi-li-um u-ni-cum me-um, dul-cis-si-mum a-mo — — rem me-

— um. Quis da-bit, quis, quis da-bit fon-tem la-cri-ma — — — —

— — — rum, quis, quis da-bit, quis, quis da-bit? —

Alessandro Della Ciaia

Alessandro Della Ciaia

8. Quid faciam misera?

Dialogus à 5 (3 voces, 2 Viole da braccio) *

Kaspar Förster Jr.
Ms. Uppsala, Univ. Library

* See Part One, pp. 135 f.

Kaspar Förster, Jr.

- ta, a quo au - xi - li - um pe - tam, co - an - gu - sta -

- - - ta?

[Amica]

Ac - ce - le - ra ad de-si-de-ra-tum cun - ctis gen-ti-bus.

Pro - pe - ra, pro - pe - ra ad___ Je - sum, pro - pe - ra

Kaspar Förster, Jr.

o - ves per - di - tas   do - mus Is - ra - e - lis.

[Mulier]

Do - mi - ne, ad - ju - va

Kaspar Förster, Jr.

9. O quam suave

Dialogus inter Christum, Magdalenam et Martham *

Benedictus a Sancto Josepho,
Encomia, op. 6 (1683)

* See Part One, pp. 154 f.

10. Dialogus inter Christum et homines (H417)

à 4 voix, 2 flûtes & 2 violons *

Prélude

Marc-Antoine Charpentier (H417)

* See Part One, pp. 170 f.

LIST OF DIALOGUES MENTIONED
OR DISCUSSED IN THIS STUDY

THIS list includes, apart from genuine dialogues, a few motets with dialogued sections.
The following abbreviations are used:

aut.   autograph
ME   modern edition
RISM   *Répertoire International des Sources Musicales*
Smither *SDD*   *Antecedents of the Oratorio: Sacred Dramatic Dialogues, 1600–1630*, ed.
H. E. Smither (Concentus Musicus, 7; Laaber, 1985)

SELECT BIBLIOGRAPHY

ADRIO, A., *Die Anfänge des geistlichen Konzerts* (Berlin, 1935).

AGAZZARI, A., *Del sonare sopra 'l basso con tutti li stromenti e dell'uso loro nel conserto* (Siena, 1607; repr. 1939, 1969).

ALALEONA, D., *Storia dell'oratorio musicale in Italia* (2nd edn., Milan, 1945); original title: *Studi su la storia dell'oratorio in Italia* (Turin, 1908).

ANTHONY, J. R., *French Baroque Music from Beaujoyeulx to Rameau* (London, 1973; rev. edn., New York, 1979).

ASTON, P., 'George Jeffreys', *Musical Times*, 110 (1969), 772 ff.

——'George Jeffreys and the English Baroque', diss. (University of York, 1970).

——'Tradition and Experiment in the Devotional Music of George Jeffreys', *Proceedings of the Royal Musical Association*, 99 (1972–3), 105–15.

AULD, L. E., *The Lyric of Pierre Perrin* (3 vols.; Henryville, Ottawa, and Binningen, 1986).

BAAB, J. C., 'The Sacred Latin Works of Kaspar Förster (1616–73)', diss. (University of North Carolina, 1970).

BANCHIERI, A., *L'organo suonario*, Op. 25 (rev. edn., Venice, 1611).

BEVERIDGE, L. P., 'Giacomo Carissimi: A Study of his Life and his Music with Latin Texts', diss. (Harvard University, 1944).

BEZE, Th. de, *Abraham sacrifiant*, ed. K. Cameron, K. M. Hall, and F. Higman (Geneva and Paris, 1967).

BOHN, E., *Die musikalische Handschriften des 16. und 17. Jahrhunderts in der Stadtbibliothek zu Breslau* (Breslau, 1890).

BROSSARD, S. DE, 'Catalogue des livres de musique, théorique et prattique, vocalle et instrumentalle', Paris, Bibl. Nat., MS Rés. Vm^8 21.

——*Dictionaire de Musique* (2nd edn., Paris, 1705; repr. 1965).

BUELOW, G., 'Rhetoric and Music', *New Grove* (London, 1980), xv. 793–803.

BUKOFZER, M., *Music in the Baroque Era* (New York, 1947).

CARRERAS y BULBENA, J. R., *El oratorio musical desde su origin hasta nuestros días* (Barcelona, 1906).

CESSAC, C., *Marc-Antoine Charpentier* (Paris, 1988).

CHAUVIN, R., 'Six Gospel Dialogues for the Offertory by Lorenzo Ratti', *Analecta musicologica*, 9 (1970), 64 ff.

CRUSSARD, C., *Un musicien français oublié: Marc-Antoine Charpentier* (Paris, 1945).

CULLEY, T. D., *Jesuits and Music*, i (Rome, 1970).

DAMMAN, R., 'Geschichte der Begriffsbestimmung Motette', *Archiv für Musikwissenschaft*, 16 (1959), 337–77.

DIXON, G., *Carissimi* (Oxford, 1986).

——'Liturgical Music in Rome, 1605–45', diss. (2 vols.; University of Durham, 1982).

——'Oratorio o motetto? Alcune riflessioni sulla classificazione della musica sacra del seicento', *Nuova rivista musicale italiana*, 17 (1983), 203–22.

——'Progressive Tendencies in the Roman Motet during the Early Seventeenth Century', *Acta musicologica*, 53 (1981), 105–19.

DONÀ, M., 'Grossi, Giovanni Antonio', *New Grove* (London, 1980), vii. 743.

Doni, G. B., *Compendio del trattato de' generi e de' modi della musica* (Rome, 1635).

Elrath, H. T., 'A Study of the Motets of Ignazio Donati', diss. (University of Rochester, 1967).

Feicht, H., '"Audite mortales" Bartłomija Pękiela', *Kwartalnik Muzyczny*, 4 (1929).

——'Muzyka w okresie polskiego baroku' (Music in the Polish Baroque Period), in *Z dziejów polskiej kultury muzycznej* (From the History of Polish Musical Culture) (Cracow, 1958).

Fogaccia, P., *Giovanni Legrenzi* (Bergamo, 1954).

Ghizlione, N., 'Giovanni Antonio Grossi: Un fecondo musicista del seicento' in *La musica sacra in Lombardia nella prima metà del seicento* (Como, 1988), 261–8.

Gibelli, V., 'La raccolta del Lucino (1608) e lo stile concertante in Lombardia' in *La musica sacra in Lombardia nella prima metà del seicento* (Como, 1988), 63–77.

Grusnick, B., 'Die Dübensammlung: Ein Versuch ihrer chronologischen Ordnung', *Svensk Tidskrift för Musikforskning*, xlvi (1964), 27–82, and xlviii (1966), 63–186.

Hitchcock, H. W., *Les œuvres de/The Works of Marc-Antoine Charpentier: Catalogue raisonné* (Paris, 1982).

——*Marc-Antoine Charpentier* (Oxford, 1990).

——'The Latin Oratorios of Marc-Antoine Charpentier', *Musical Quarterly*, 41 (1955), 41–65.

Hudemann, O., 'Die protestantische Dialogkomposition im 17. Jahrhundert', diss. (Universität Kiel, 1941).

Jachimecki, Z., *Wpływy włoskie w muzyce polskiej, Cz. I: 1540–1640* (Italian Influences in Polish Music, Part I: 1540–1640) (Cracow, 1911).

Jones, A. V., *The Motets of Carissimi* (2 vols.; Ann Arbor, Mich., 1982).

Kircher, A., *Musurgia universalis* (2 vols.; Rome, 1650; repr. 1969).

Kroyer, T., 'Dialog und Echo in der alten Chormusik', *Jahrbuch der Musikbibliothek Peters 1909*, 13 ff.

Kurtzmann, J., 'Giovanni Francesco Capello, an Avant-gardist of the Early Seventeenth Century', *Musica disciplina*, 31 (1977), 155–82.

Launay, D., 'Guillaume Bouzignac', *Musique et liturgie*, 21 (1951), May–June.

——Introduction to *Anthologie du motet polyphonique en France (1609–1661)* (Paris, 1963).

Le Cerf de la Viéville, J. L., *Comparaison de la musique italienne et de la musique française* (Brussels, 1704–6; repr. 1972; 2nd edn. in Pierre Bourdelot et Jacques Bonnet, *Histoire de la musique et de ses effets* (Amsterdam, 1725; repr. 1966)).

Lowe, R. W., *Marc-Antoine Charpentier et l'opéra de collège* (Paris, 1966).

MacDonald, J. A., 'The Sacred Vocal Music of Giovanni Legrenzi', diss. (University of Michigan, 1964).

Massenkeil, G., 'Carissimi, Giacomo', *New Grove* (London, 1980), iii. 785–94.

——'Die Wiederholungsfiguren in den Oratorien Giacomo Carissimis', *Archiv für Musikwissenschaft*, 13 (1956), 41–60.

——'Zu einigen dialogischen Concerti des frühen 17. Jahrhunderts', in *Beiträge zur Geschichte des Konzerts: Festschrift Siegfried Kross zum 60. Geburtstag* (Bonn, 1990), 13–19.

Mattheson, J., *Grundlage einer Ehren-Pforte* (Hamburg, 1740; modern edn. by M. Schneider, Berlin, 1910, repr. 1969).

MEER, J. VAN DER, 'Benedictus a Sancto Josepho vom Karmeliterorden', *Kirchenmusikalisches Jahrbuch*, xlvi (1962), 99–120, and xlvii (1963), 123–4.

MERSENNE, M., *Harmonie universelle* (Paris, 1636–7; repr. 1963).

MEYER, K., 'Das Offizium und seine Beziehung zum Oratorium', *Archiv für Musikwissenschaft*, 3 (1921), 371–404.

MORTOFT, F., *His Book Being his Travels Through France and Italy, 1658–1659*, ed. M. Letts (London, 1925).

MOSER, H. J., *Die mehrstimmige Vertonung des Evangeliums* (Leipzig, 1931; repr. 1968).

MYERS, P. A., 'Antonelli, Abundio', *New Grove* (London, 1980), i. 491–2.

NOACK, R., 'Dialog', *Die Musik in Geschichte und Gegenwart*, iii (Kassel-Basel, 1954), cols. 301 ff.

NOSKE, F. R., *Music Bridging Divided Religions: The Motet in the Seventeenth-Century Dutch Republic* (2 vols.; Wilhelmshaven, 1989).

——'Observations on the Seventeenth-Century Latin Dialogue' (text in Russian), *Early Music in the Context of the Contemporary Culture: Problems in Interpretation and Source Study. Musicological Congress, Moscow Conservatory* (Moscow, 1989), 397–410.

——'Sacred Music as Miniature Drama: Two Dialogues by Carlo Donato Cossoni (1623–1700)', in *Festschrift Rudolf Bockholdt zum 60. Geburtstag* (Pfaffenhofen, 1990), 161–81.

——'Sul dialogo latino del seicento: Observazioni', *Rivista italiana di musicologia*, 24 (1989), 330–46.

OSTHOFF, H., 'Domenico Mazzocchis Virgil-Kompositionen', in *Festschrift Karl Gustav Fellerer* (Regensburg, 1962), 407 ff.

PASQUETTI, G., *L'oratorio musicale in Italia* (Florence, 1906; 2nd edn., 1914).

PRZYBYSZEWSKA-JARMIŃSKA, B., 'Kacper Förster Junior, Zarys biografii', *Muzyka*, 3 (1987), 3–19.

QUITTARD, H., *Henry Du Mont: Un musicien français au XVIIe siècle* (Paris, 1906).

——'Un musicien français oublié du XVIIème siècle français: Guillaume Bouzignac', *Sammelbände der internationalen Musikgesellschaft*, vi (1904–5), 356 ff.

RANUM, P., 'A Sweet Servitude: A Musician's Life at the Court of Mlle de Guise', *Early Music*, 15 (1987), 347–60.

ROCHE, J., 'Cross-currents in Milanese Church Music in the 1640s: Giorgio Rolla's Anthology "Teatro Musicale"', in *La musica sacra in Lombardia nella prima metà del seicento* (Como, 1988), 14–29.

——'Grandi, Alessandro', *New Grove* (London, 1980), vii. 631–4.

——*North Italian Church Music in the Age of Monteverdi* (Oxford, 1984).

——'The Duet in Early Seventeenth-Century Italian Church Music', *Proceedings of the Royal Musical Association*, 93 (1966–7), 33–5.

ROSENTHAL, K., 'Steffano Bernardis Kirchenwerke', *Studien zur Musikwissenschaft*, 15 (1928), 40–61.

SCACCHI, M., *Cribum musicum ad triticum Syferticum, seu Examinatio succinta psalmorum* (Venice, 1643).

SCARPETTA, U., 'Michaelangelo Grancino, maestro di capella del Duomo di Milano', in *La musica sacra in Lombardia nella prima metà del seicento* (Como, 1988), 247–57.

SCHAAL, R., *Die Musikhandschriften des Ansbacher Inventars von 1686* (Wilhelmshaven, 1966).

SCHERING, A., *Geschichte der Musik in Beispielen* (Leipzig, 1931).

——*Geschichte des Oratoriums* (Leipzig, 1911; repr. 1970).

——Review of D. Alaleona, *Studi su la storia dell'oratorio in Italia* (Turin, 1908), in *Zeitschrift der internationalen Musikgesellschaft*, x (1908), 178 ff.

——Review of G. Pasquetti, *L'oratorio musicale in Italia* (Florence, 1906), in *Zeitschrift der internationalen Musikgesellschaft*, ix (1907), 44–6.

——Review of J. R. Carreras y Bulbena, *El oratorio musical* (Barcelona, 1906), in *Zeitschrift der internationalen Musikgesellschaft*, viii (1906), 200.

SEELKOPF, M. R., 'Das geistliche Schaffen von Alessandro Grandi', diss. (Universität Würzburg, 1973).

SEIFFERT, M., Introduction to J. P. Kriegcr, *21 ausgewählte Kirchenkompositionen* (Denkmäler deutscher Tonkunst, 53; Leipzig, 1916).

SHIGIHARA, S., *Bonifazio Graziani (1604/5–1664): Biographie, Werkverzeichnis und Untersuchungen zu den Solomotetten* (Bonn, 1984).

SMALLMAN, B., 'Endor Revisited: English Biblical Dialogues of the Seventeenth Century', *Music and Letters*, 46 (1965), 137 ff.

SMITHER, H. E., 'Carissimi's Latin Oratorios: Their Terminology, Functions, and Position in Oratorio History', *Analecta musicologica*, 17 (1976), 54–78.

——*A History of the Oratorio*, i, ii (Chapel Hill, 1977).

——'Romano Micheli's "Dialogus Annuntiationis" (1625): A Twenty-Voice Canon with Thirty "Obblighi"', *Analecta musicologica*, 5 (1968), 34 ff.

——'The Latin Dramatic Dialogue and the Nascent Oratorio', *Journal of the American Musicological Society*, 20 (1967), 403–33.

——'What is an Oratorio in Mid-Seventeenth-Century Italy?', in *International Musicological Society Congress Report*, xi (Copenhagen, 1974), 657–63.

SOLERTI, A., *Le origini del melodramma* (Turin, 1903; repr. 1969).

SPAGNA, A., *Oratorii, overo melodrammi sacre*, ii (Rome, 1706).

SPELLERS, S., 'Collegium Musicum te Groningen', in *Bouwsteenen: Derde jaarboek der Vereeniging voor Noord-Nederlands Muziekgeschiedenis* (n.p., 1881), 22–9.

SPINK, I., 'English Seventeenth-Century Dialogues', *Music and Letters*, 38 (1957), 155 ff.

STEFANI, G., *Musica e religione nell'Italia barocca* (Palermo, 1975).

SZWEYKOWSKI, Z., Introduction and edition of B. Pękiel, 'Audite mortales' (Wydawnictwo dawnej muzyki polskiej, Cracow, 1968).

WHENHAM, J., 'Dialogue, Sacred', *New Grove* (London, 1980), v. 419–20.

——*Duet and Dialogue in the Age of Monteverdi* (Ann Arbor, Mich., 1982).

WINTER, C., 'Studien zur Frühgeschichte des lateinischen Oratoriums', *Kirchenmusikalisches Jahrbuch*, xlii (1958), 64–76.

WINTERFIELD, C. von, *Johannes Gabrieli und sein Zeitalter* (2 vols.; Berlin, 1834; repr. 1965).

WITZENMANN, W., 'Autographe Marco Marazzolis in der Biblioteca vaticana', *Analecta musicologica*, 7 (1969), 36–86, and 9 (1970), 203–94.

——*Domenico Mazzocchi, 1592–1665: Dokumente und Interpretationen (Analecta musicologica*, 8 (1970)).

INDEX OF NAMES